DELESSERIA SA. GUINEA.

Drawn & Etched lith J A Patten Co. sc

Published by John Van Voorst, Paternoster Row 1851

THE

SEA-WEED COLLECTOR'S
GUIDE:

CONTAINING

Plain Instructions for Collecting and Preserving,

AND A LIST OF ALL THE

KNOWN SPECIES AND LOCALITIES

IN

GREAT BRITAIN.

By J. COCKS, M.D.,

DEVONPORT.

"Seek for the beautiful in every cave,
On rocks uncovered by retreating waves,
————— Where the Sea-Weed grow."

LONDON:

JOHN VAN VOORST, 1, PATERNOSTER ROW.

M.DCCC.LIII.

THE

SEA-WEED COLLECTOR'S
GUIDE:

CONTAINING

Plain Instructions for Collecting and Preserving,

AND A LIST OF ALL THE

KNOWN SPECIES AND LOCALITIES

IN

GREAT BRITAIN.

By J. COCKS, M.D.,

DEVONPORT.

"Seek for the beautiful in oozy cave,
On rocks uncovered by receding waves,
—— Where the Sea Weeds grow."

LONDON:

JOHN VAN VOORST, 1, PATERNOSTER ROW.

M.DCCC.LIII.

TO

LIEUT. COLONEL C. HAMILTON SMITH,

K.H., K.W., F.R.S., F.L.S.,

&c., &c.,

PRESIDENT OF THE DEVON AND CORNWALL

NATURAL-HISTORY SOCIETY,

HONORARY PRESIDENT OF THE PLYMOUTH INSTITUTION,

WHOSE UNWEARIED EXERTIONS IN
THE PURSUIT, AND IN ADVANCING THE SCIENCE, OF
NATURAL HISTORY,
FOR UPWARDS OF HALF A CENTURY,
HAS BEEN DULY ACKNOWLEDGED AND APPRECIATED
THROUGHOUT THE CIVILIZED WORLD,

THIS LITTLE WORK

IS, BY PERMISSION, DEDICATED, BY HIS

VERY FAITHFUL

AND OBLIGED SERVANT,

JOHN COCKS, M.D.

INTRODUCTION.

It has been alleged by students, and certainly with much truth, that the instructions for collecting and preserving sea-weeds, to be found in recently-published works, are not sufficiently comprehensive and explicit; and that other and more practical details are required to enable them to exhibit in a satisfactory manner the specimens they have been at the trouble of collecting.

It is not my intention to offer any apology for publishing a work like this, but to content myself with stating, that for several years I have been solicited to do so, by numerous persons residing in my own neighbourhood, and by correspondents and strangers from various quarters of the kingdom, and the sister island.

More recently these applications have been supported by the recommendation of several scientific algologists; and I have at length yielded to the general desire, and in the following pages have given publicity to the method I have myself for many years adopted, and have found to be most successful.

Specimens laid out by myself, in the manner hereafter to be described, have been repeatedly spoken of in terms of high commendation by Sir William Hooker, Professor Harvey, Mrs. Griffiths, and other algologists.*

This is not, however, the only object I have had in view. I have, in addition, compiled a list of all the marine Algæ that are at present known and recognized as British, including even the latest discoveries, as described in Dr. Harvey's beautiful work, the 'Phycologia Bri-

* At the Annual Meeting of the Royal Cornwall Polytechnic Institution, Falmouth, held Sept., 1852, the author was awarded the silver medal, being the first-class prize in Natural History, viz., for three vols., imperial folio, containing specimens of marine Algæ.

tannica,' the systematic arrangement of which I have taken as my guide.

I have also enumerated the respective localities where the different species of Algæ may be found, together with the time of their appearance. This information cannot be otherwise than useful to the young algologist, as it will not only save him a great deal of unnecessary trouble, but will enable him, when in search of a particular plant, at once to direct his steps to the precise spot, which he might not otherwise be enabled to discover.

It is not intended, neither is it possible, within the confined limits of a work like this, to enter into a description of the character and peculiar structure of the marine Algæ. It contains, as its title expresses, with the exception of a few general remarks, and the list of marine Algæ previously adverted to, familiar instructions for collecting, laying down, and preserving sea-weeds.

The author flatters himself that the plain information here given, will be found amply sufficient for this purpose; and whilst the young algologist continues to prosecute his

researches, he will feel himself amply rewarded
for his trouble : and. the author further hopes
that the fortunate discovery of a plant hitherto
unknown and undescribed, will stimulate him
to the acquisition of information of a higher
and more scientific character, and progres-
sively lead him to the perfect study of this
exceedingly beautiful order of Cryptogamia—
flowerless plants.* In this pursuit he will be
greatly assisted by the perusal of Dr. Harvey's
excellent work, the 'Manual of British Marine
Algæ,'† of which a second edition, with illustra-
tions, has lately been published.

* The author is the fortunate discoverer of that exceedingly rare plant, Stenogramme interrupta, thus noticed by
Dr. Harvey :—

" This very interesting plant, by far the most important
addition which has been made to the British Marine Flora,
since the commencement of the present work, was discovered on the 21st October, 1847, by Dr. John Cocks, of
Plymouth, among rejectamenta on the shore of Bovisand,"
&c., &c.—*Phycologia Britannica*, vol. ii.

† The author here begs to acknowledge that, in the
course of this little work, he has taken the liberty of making
many extracts, both from Dr. Harvey's 'Manual,' and from
the ' Phycologia Britannica.'

Those also who, happily unembarrassed by the sterner realities of the business of life, and who, in the summer season, are attracted to the sea-shore in search of health or recreation, may, under the guidance of this little work, sweeten the vacuity of unemployed hours with healthful exertion, and enrich their cabinets, as well as their friends, with specimens of most exquisite and natural beauty.

JOHN COCKS, M.D.

PREFACE.

It has long been admitted that there are few studies better calculated to expand the mind, and gratify a laudable thirst after knowledge, than the study of Natural History; and, taking it for granted that all knowledge is pleasure, as well as power, it necessarily follows that the pleasure derived, and the power obtained, will be in direct proportion to the labour bestowed.

Whilst various other branches of science are cultivated to improve the reasoning faculties of the mind, their influence on the heart and affections are comparatively feeble.

Natural History has a different effect, and may be said to improve and humanise the whole man; whilst the pleasure and instruction inseparable from its study, whenever it is pursued in a truly philosophical spirit, is very great.

Moreover, a comprehensive knowledge of Natural History cannot be acquired without producing a corresponding improvement in taste, literature, and the fine arts.

If that only which is true can be of permanent utility in our social condition, surely the study of natural forms, and their structure, must be to the historian, the sculptor, the painter, and the poet, a means of acquiring new and correct information, and therefore a more perfect delineation of their united labours. This, reacting upon the public mind, will lift it to a more perfect appreciation of the beauties of the fine arts, and the value of scientific literature.

Amongst the reasons assigned as tending to discourage a desire for acquiring instruction and information in the various branches of Natural History, complaints have been made, and with much truth, against many of the works which have been published on these subjects, of their being written in language which would appear as designed rather to display the learning of the writer, than to state the facts which such learning ought to convey; whilst, on the other hand, it would be more desirable, as far as it is practicable, instead of using scientific language and terms, to condescend to the use of such as shall make the meaning intelligible to the general reader.

There is also another cause which has greatly impeded the study of Natural History, namely, the difficulty experienced in obtaining access to the more expensive works.

This objection, however, does not apply so much to works on Botany, as many highly-useful publications are attainable, and in a very inexpensive form. This has, in all probability, induced many persons to follow so fascinating a branch of Natural History, in preference to any other.

Latterly, the science of Algology has attracted to its interesting pursuit numerous votaries, not only from amongst those who reside along the coasts of the kingdom, celebrated for the production of Algæ, those beautiful symbols

of the munificence of the Creator,—who, when
he clothed the valleys and the mountain-
sides with verdure, forgot not to fringe the
caverns of the deep with plants of lovely form,
and varied hue,—but also from among those
who, in the midst of crowded cities, can turn
with a pure taste from the pursuits and associ-
ations of common life, to the study of that great
book in which the philosophic mind will always
recognize and admire the distinct tracery of the
hand of the great Creator of all things.

Contemporaneously, too, with this popular
display of interest, there have issued from the
press several inexpensive and useful publications, so constructed and arranged as to assist
the uninitiated student in his earlier essays.

Even those who do not pretend to the study
of the marine Algæ in a scientific spirit, will
derive much pleasure from the collection and
preservation of the numerous varied and interesting specimens which are so beautifully con-
spicuous when neatly laid out upon paper, and
arranged with taste.

Nor need the young student who aspires to
become an algologist, doubt that he will be

successful in his efforts, and obtain his full share of the superior pleasure which scientific knowledge can so liberally bestow : he will find that every onward step will render those which are to follow less difficult ; and every increase of knowledge thus gained, will be to him a new source of delight, and a stimulating inducement to proceed.

The Sea-weed Collector's Guide.

CHAPTER I.

THERE is no country, I believe, that has so great a variety of marine Algæ as Great Britain. The shores contiguous to Plymouth are peculiarly rich, both as regards the number and rarity of species.

In colour, the Algæ exhibit three principal varieties, with, of course, numerous intermediate shades, namely, grass-green (Chlorospermeæ), olivaceous (Melanospermeæ), and red (Rhodospermeæ).

The *grass-green* colour is characteristic of those found in fresh water, or in very shallow parts of the sea, along the shores, generally above half-tide level, and is rarely seen in those plants which grow at any great depth. But to this rule there are exceptions, sufficiently numerous to forbid our assigning the prevalence of this colour altogether to shallowness of water.

B

Several of the most perfect Confervæ and Siphoniæ grow beyond the reach of ordinary tides; and others are sometimes dredged from very considerable depths. The greater mass, however, of the green-coloured series are considerably submerged.

The *olivaceous brown*, or olive-green series, is almost entirely confined to marine species, and is, in the main, characteristic of those that grow at half-tide level, becoming less frequent towards low-water mark; but it frequently occurs also at greater depths, in which case it is very dark, and passes into brown, or almost black.

The *red* series, also, is almost exclusively marine, and reaches its maximum in deep water, though some varieties occur at half-tide level. When above this, they assume either purple, orange, or yellow tints, and sometimes even a cast of green. They are, however, very rarely pure red much within the range of extreme low-water mark, higher than which many of the more delicate species will not vegetate; and those that do exist, degenerate in form, as well as in colour, as they recede from it. How far below low-water mark the red species extend, has not been ascertained; but those from the extremest depths of the sea that we are acquainted with, are of the olive series in its darkest form.

In the ' Manual of British Algæ,' published by Dr. Harvey, he says :—" Among plants in general nothing is so variable or uncertain as colour. Far from serving as a mark to distinguish groupes, or genera, the utmost to which it can pretend is to separate one variety of species from another.

" Among Algæ, on the contrary, it has been ascertained that the classes of colour enumerated before, are, to a great extent, indicative of structure, and of natural affinity. Thus, the green species are of the simplest structure, and differ remarkably in their mode of reproduction from either of the other tribes :—their seeds being endowed at the germination with a sort of motion which some have called voluntary, but which really does not possess that animal property.

" The olivaceous (Melanospermeæ) are the most perfect and compound, and reach the largest size. The red series (Rhodospermeæ) form a group not less distinguished by the beauty and delicacy of their tissue, than by producing seeds under two forms ; — thus possessing what is called a double fructification."

Most Algæ are, at some period of their growth, found attached to other substances, by means of a root, or, at least, a hold-fast. It has been doubted whether, as no distinct vessels of absorption have been discovered, they receive any nourishment through this organ ; but the question is by no means settled.* Thus much is, at

* From attentive observation, during the last four or five years, I have myself come to the conclusion that marine

least, certain :—they appear to be as much influenced as other plants by the soil in which they grow; for different species of rock afford different kinds in greater perfection than others, and a large number of those that are parasitical confine themselves to particular species; a fact that should not fail to be recollected by the young algologist. The selection of habitat, it has been argued, would seem to prove that the root is not so sluggish an organ as it has been supposed to be. It does not, however, present much modification, and rarely attains a large size, differing from that of most land-plants, in which, as they grow in size and height, the root also extends, and increases in proportion.

The cellular tissue of Algæ presents some varieties. The most common form of the cellule is cylindrical, often of very small diameter in proportion to its length; and in such cases the cellules always cohere, by the ends, into threads, or filaments, bundles of which, either branched or simple, form the frond,* by lateral cohesion.

Algæ do not derive any nourishment whatever from the roots, and that these organs serve merely as hold-fasts. I have frequently gathered as fine specimens growing on the mooring buoys in Plymouth Harbour, covered thickly with pitch as well as copper, as in other places; neither of which substances would, I presume, be likely to afford nutritive matter for promoting vegetation.

* The term "frond," when applied to a sea-weed,

The fronds of many of the simple kinds, as Confervæ, &c., consist of a single thread, or string of cellules, or joints. Those which are more compound may generally be resolved into such threads by macerating small portions either in hot water, or, if that prove ineffectual, in diluted muriatic acid.

In the fructification, we find many modifications of structure without much real difference, either in the manner in which the fruit is perfected, or in the seed that is produced.

The seeds that are finally formed in all the tribes of genuine Algæ, appear pretty nearly to agree in structure, and to consist of a single cellule, or bag of membrane, filled with a very dense, dark-coloured, granular, semi-fluid mass, called the endochrome. These seeds, on germination, produce perfect plants, resembling that from which they sprung. Nothing at all resembling floral organs has been noticed in any ; and all that we know of the fructification is, that it takes place with regularity,—arising from the same parts of the frond, and having the same appearance in plants of the same kind. Its growth may be watched from the commence-

---

signifies every part of the plant, with the exception of the root, and occasionally of the stem, if well developed, and distinct from the other portions of the plant not included under this term.

ment, when what we may call the ovule, or germ of the future seed, begins to swell. But nothing whatever has been ascertained that throws the smallest light on the process of fecundation.

Many Algæ, perhaps all the red series, are furnished with a double system of fructification, called primary and secondary fruit,—terms which are given for convenient distinction, without intending them to mean that one is of more or less importance than the other, for the seed formed in each is equally capable of producing a new plant. What is called primary, is generally placed in capsules, which are either globose or pitcher-shaped, or, at least, a large number of seeds are collected into compact sphærical clusters, and immersed in the frond. In the secondary, on the contrary, the seeds, which are commonly called granules, are usually placed in cloud-like or defined patches, called *sori*, or in distinct portions of the frond. But in many genera, as in Odonthalia, Dasya, Griffithsia, &c., proper receptacles, of various shapes, are formed for their reception.

There is, besides, a really anomalous structure, connected with an imperfect attempt at fructification, not uncommonly found on several Floridiæ. This, to which Agardh gives the name of *nemathecium*, is a wart-like protuberance, of a very irregular figure, and generally large size, consisting entirely of concentric filaments, with coloured joints, in all respects resembling those that form the periphery, but much longer.

To the naked eye, these warts often resemble capsules, and as such have been frequently described ; but they never contain any seeds.

Another anomalous body, simulating fruit, frequently occurs in some of the filamentous tribes, especially in the genus Polysiphonia, and in the species P. fibrata, P. fibrillosa, and P. fastigiata, to which Agardh has given the name of *antheridium*. It is a very minute pod-like or lanceolate body, of a yellow colour, containing a granular fluid, borne on the colourless, long-jointed fibres that are found, at particular seasons, issuing from the tips of the branches in several, if not all, of the Polysiphoniæ. The nature of these minute organs, it is thought by Dr. Harvey, deserves more attention than it has obtained ; for he says, " I am strongly of opinion that they are produced with too much regularity to be regarded as accidental."

After the foregoing observations, chiefly taken from Dr. Harvey's Manual, and which are introduced with the intention of affording to the young algologist some useful information of a general character, I now proceed to the practical part of the subject, by endeavouring to explain, as clearly as I am able, the best manner of collecting marine Algæ, and of afterwards preserving them. Having already stated that this has been a favourite pursuit with me for many years, and having tried various modes, I have found, after a good deal of experience and practice, that the method I am about to describe is preferable to any other.

CHAPTER II.

WHEN collecting sea-weeds, it is advisable to gather them either from pools left by the receding tide, where many species of them grow, or in places where they have been recently thrown up by the sea; for, after they have been exposed to the sun and air, and more especially during the summer months, they become decomposed, and lose their natural colour and appear-ance.* In gathering your plants, endeavour, if possible, to preserve the root, as it makes the specimen more perfect, and, in consequence, of more value; and also contrive to select, in preference to others, plants that are in fruit, for these are more highly estimated by scientific algologists than such as are bar-

* All Algæ growing within the limits of the tidal influence, are to be sought at low water, especially the lowest water of spring-tides, for many of the rare and more interesting kinds are found only at the verge of low-water mark, either along the margin of rocks partially laid bare, or, more frequently, fringing the deep tide-pools left, at low water, on a flattish shore. The northern, or shaded, side of the pool will be found richest in *red* Algæ; and the most sunny, in those of an *olive* or *green* colour.

ren.*   Immediately after gales of wind, and more particularly after those occurring a little before or during the
spring-tides, which take place at the period of every new
and full moon, the shore should be diligently explored,
and the rejectamenta thrown up by the sea carefully
turned over and examined.†   The action of the sea dur-

---

* In gathering Algæ from their native places, the whole
plant should be plucked from the very base; and if there
be an obvious root it should be left attached.   Young col-
lectors are apt to pluck branches or mere scraps of the
larger Algæ, which in most cases afford no just notion of
the mode of growth or natural habit of the plant from which
they have been snatched, and are often insufficient for the
first purpose of a *specimen*, that of ascertaining the plant to
which it belongs.

In many of the leafy fucoid plants, such as Sargassum,
&c., the leaves that grow on the lower and on the upper
branches are quite different; and were a lower and an
upper branch picked from the same root, they might be
found so dissimilar as to pass for portions of different species.   It is therefore very necessary to gather, when it can
be done, the whole plant, including the root.   It is quite
true that the large kinds may be judiciously divided; but
the young collector had better aim at selecting *moderately-
sized* specimens of the entire plant, than attempt the divi-
sion of large specimens, unless he keep in view this maxim:
— every botanical specimen should be an epitome of the
essential marks of a species.

† In selecting from rejectamenta cast upon the shore,
we should take those specimens only that have suffered
least in colour or texture, by exposure to the air.

ing heavy gales of wind has the effect of loosening, and detaching from their place of growth, many species, and more especially those growing in deep water, that are advanced to maturity, which are soon afterwards washed ashore, and will generally repay the algologist for his trouble in searching for them. He ought, indeed, never to miss so favourable an opportunity of adding to his collection. Moreover, he will frequently meet with some of the rarer species which he could not otherwise obtain, except by dredging, a rather troublesome operation, and one which must always be regarded as attended with very doubtful success.

Although all parts of the strand should be diligently examined, yet the collector should direct his attention especially to three different points, *viz.*, high-water mark, low-water mark, and half-tide level, for in these places he will find the largest masses of sea-weeds accumulated.*

Many species are found which flourish and vegetate more luxuriantly in localities where small rills of fresh water run into the sea.

---

* To persons collecting at Plymouth, it may be desirable for them to know that they will find the largest quantity of rejectamenta thrown up at the following localities, *viz.*, under the *Hoe*, on the mud-bank of Cremell Passage, at Redding Point, opposite the ballast-pond at Tor Point, and on the shore leading from Bovisand down as far as the Rennie rocks.

Wherever there are *tide-mills* in the neighbourhood, they should occasionally be visited. The mill-dam, the sluices, and wherever there is running water, will always afford many good varieties ; and the collector should bear in mind that these are best attained at neap-tides.

Although, as has been before stated, a great number of species are occasionally to be met with, scattered on the shore, which have been thrown up by the sea ; yet the zealous algologist, who aims at possessing a complete and superior collection, will scarcely rest satisfied with these methods only of obtaining his supplies.

As he finds his stock gradually accumulating, he will be stimulated to make further efforts to add to it ; and particularly when, on looking over the list of species, he is struck with the disagreeable fact, that there is still a great number deficient. During the spring tides, in the course of his exploratory rambles, he will observe, even at extreme low-water mark, many beautiful varieties, growing either in deep pools of water or in places quite out of reach ; and it will occasionally happen that these are rare species, or such as he is much in want of. In order to be prepared for such an exigency, I generally carry with me a stout walking-stalk, to the end of which, when I arrive at the water-side, I screw on an instrument somewhat resembling that which gardeners are in the habit of using for cutting up weeds. By this means, I am enabled to detach and secure many plants growing on the sides of perpendicular rocks, or in deep pools of water, which otherwise I could not obtain ; whilst the stick, thus armed, serves also as a help and support, and often saves the collector from a tumble when scrambling over the slippery rocks. But, even with this auxiliary, he will find that there are many plants not attainable when approaching their place of growth from the shore. Here the only alternative is to repair to the locality in a boat. He will, indeed, find it greatly to his advantage to hire a boat occasionally ; and, with the assistance of an obliging boatman, he will, in the course of a few hours, in fine weather, and when the sea is smooth, be able, with very little difficulty, to obtain not only an abundant supply of plants, in excellent condition, but he will have an opportunity of observing the places where particular species grow, and the nature of the substances they are attached to. Some he will perceive growing only on the perpendicular sides of rocks ; whilst others will be found on their flat surfaces. Many he will discover growing parasitically on the different varieties of Fuci, and on the stems of the Laminariæ. One, a very interesting species, Microcladia glandulosa, is generally found growing on the fronds of Rhodomenia laciniata, and on Nitophyllum laceratum.

For the more zealous and enterprizing algologist another plan for procuring specimens remains to be men-

tioned ; and, as several very rare species are seldom to
be procured by any other means, this should on no account be overlooked.   This is by *dredging*, which,
although a rather troublesome and dirty operation, will
often repay him for his exertions.   Many plants which
grow only in very deep water, are scarcely ever found
in good condition, except by the process of dredging.
There is also another great advantage attending the
use of the dredge, for, on carefully examining the con-
tents brought up from the bottom of the sea, the collector will occasionally find scarce shells, corallines and
zoophytes, besides many species of Crustacea, &c.

After gathering our plants, the next point to be attended to is to give them a good washing before leaving
the shore, either in the sea or in some of the pools left
by the receding tide, removing from them, at the same
time, as far as may be practicable, all fragments of decayed sea-weeds, and other extraneous bodies, such as
particles of sand and gravel, portions of the softened
surface of sandstone or argillaceous rock on which the
plants have been growing, together with the smaller
Testacea, &c., &c.

When gathering our plants, however, it cannot be
expected we can devote any considerable portion of time
to this cleansing, our principal object being to collect,
and remembering that we shall be compelled to relin-
quish our pursuit soon after the tide begins to flow;

for, after all, on our return home, it will be found there is still a great deal of work to do before the specimens are in a fit state to be finally committed to paper, since foreign substances will still remain attached to them with much pertinacity even after we have supposed them to be perfectly clean.*

In carrying them home, a tin vessel, made something in shape like a lady's reticule, well painted inside with white paint, and covered externally with one or more coats of black japan varnish, will be found very convenient. For the coarser weeds, I use, myself, a fishing-basket, lined with vulcanized India rubber, or gutta percha, which I sling across my shoulder with a broad leather strap. Some prefer flat, wide-mouthed glass or zinc bottles. The finer species, such as the Callithamniæ, Griffith-siæ, &c., &c., should always be kept apart from others. Strong, flat, white glass bottles, commonly termed *toad-*

* Several duplicate specimens of every kind should, if attainable, be always preserved, and more especially where the species is a variable one. Very many Algæ vary in the comparative breadth of the leaves, and in the degree of branching of the stems; and when such varieties are noticed, a considerable series of specimens is often requisite to connect a broad and a narrow form of the same species.

A neglect of this care leads to endless mistakes in the after-work of identification of species, and has been the cause of burdening our systems with a troublesome number of synonyms.

mouthed, holding about two or three ounces each, are best suited to bring home these species.*

I must now caution young beginners to keep apart from other plants all the different varieties of the genus Desmarestia, for they possess the peculiar property of changing the colour of, and very soon decomposing, all other plants, especially those of the finer species, with which they may come in contact.

Notwithstanding what has been alleged to the contrary, most of the Desmarestiæ, if kept out of sea-water, even for a very inconsiderable time, soon become flaccid, and rapidly advance towards decay. Specimens of Desmarestia viridis, taken in the spring, which are exceedingly beautiful, if properly laid out, are very difficult to preserve, and can scarcely be brought from any great distance without being decomposed, or fast approaching to that state. It is therefore necessary to remember that these plants should always be brought home in vessels filled with sea-water, and that only a few specimens

---

* A precaution often absolutely necessary, for many of the delicate *red* Algæ rapidly decompose, if exposed, even for a short time, to the air, or if allowed to become massed together in the same bottle, as crowding encourages decomposition; and when this has once begun, it spreads with fearful rapidity As many of these Algæ will not keep, even in large vessels of sea-water, from one day to another, the sooner they are arranged for drying, the better.

should be placed together in the same bottle; and, further, when laid out, in the manner hereafter to be described, fresh water must never be used. If allowed to come in contact with fresh water, even for a very short time, it has the effect of causing the fine and delicate ramuli to become clotted together; these it is afterwards impossible to separate; and the appearance of the specimen is consequently very much injured, if not altogether spoiled.

Let us now suppose we have reached home, with the various plants collected; and, being furnished with a sufficient supply of sea-water, the first step to be taken is to examine each one separately, and carefully remove every particle of extraneous matter that may be attached. These foreign bodies are more easily detected by placing the specimen in a flat, white dish filled with salt water, remembering that the Calithamnia, and other delicate varieties, require to be first attended to; for, notwithstanding the pains we may have taken to clean our specimens beforehand, we shall often find, when they are fairly spread out, that there are still some minute particles adhering to them.

These are effectually removed with a pair of *dissecting-forceps,* which are always to be procured of any surgical-instrument maker, and are, indeed, almost indispensable in laying out marine Algæ. They will, besides, be found most useful for various purposes

C

difficult to describe, but which the operator will soon find out, as he proceeds, and learn to appreciate them accordingly.

As a preliminary step, it is scarcely necessary to observe that a sufficient quantity of paper should be previously provided, on which the plants are to be laid out, and are finally to remain. Now, the quality of the paper is a matter of considerable importance : it requires to be judiciously selected, for the purpose intended ; for it frequently happens that a great error is committed in this respect, not only by the novice, but also by the more experienced algologist, in using paper of a thin and inferior quality, which very much injures the appearance of the specimens. There are some species in particular, that contract so much in drying, as to pucker the edges of the paper if it is not sufficiently thick; and these are then seen to considerable disadvantage. After trying various sorts of paper, I now use only one kind, which, after long practice, I have found to be better adapted than any other, with the exception of drawing-paper, for the display of marine Algæ. This is what the stationers term a *thick printing demy.*

This paper, it it is proper to mention, should weigh about thirty-four or thirty-five pounds a ream, supposing it to possess the ordinary dimensions of that description of paper when folded once, *viz.*, about $17\frac{1}{2}$ inches by 11. Having chosen a paper of this kind,

I would strongly recommend that no other should be used. It is also desirable, and will be found very convenient, that a quantity of papers of different sizes should be previously cut, and in readiness for displaying the plants,—a practice which will enable the operator to dispose of his materials to the best advantage.

This observance serves, also, to give a neatness and uniformity to a collection, not to be accomplished by using papers cut at random, or of casual dimensions.

It will also be necessary to be provided with a suitable vessel, filled with clean fresh water, for floating out and spreading the specimens on paper. For this purpose, a flat, white dish, glazed within, about three inches deep, eighteen inches long, and fourteen inches wide, will be found most convenient. When required for use, it is to be filled nearly to the brim with very clean water.

Some persons are in the habit of using the shallow dishes that are commonly employed in bringing meat to table : but such vessels are not at all calculated for floating out sea-weeds ; for, unless there is a sufficient depth of water to allow the hand to pass under the paper on which the specimen is to be spread, it can never be neatly or naturally arranged.

You must also have in readiness a few quires of paper for drying your specimens after they are removed from the water. I always use two sorts, *viz.*, first, Bentall's botanical drying-paper,* and afterwards another of finer texture, such as is used by druggists, and called filtering-paper, or, what will answer equally well, thick blotting-paper. All these can be procured at any respectable stationer's. Each sheet should be folded to a quarto size, and had better remain uncut, for the convenience of being hung across a line, to dry for future use. Three or four smoothly-planed deal boards, about an inch thick, and thirteen or fourteen inches square, will also be required, as well as a good many pieces of fine linen, cambric muslin, or calico, about the same size, that which has been worn previously being preferable to new.†

* To be procured of Mr. Newman, 9, Devonshire St., Bishopsgate St., London.

† When it is the collector's object to preserve Algæ in the least troublesome manner, and in a rough state, to be afterwards laid out, and prepared for pressing at leisure, the specimens, fresh from the sea, are to be spread out, and left to dry, in an airy, but not too sunny, situation. They are, besides, not to be washed or rinsed in fresh water; nor is their natural moisture to be squeezed from them. The more loosely and thinly they are spread out, the better; and in warm weather they will be sufficiently dry after a few hours' exposure to allow of packing. In a damp state of the atmosphere, the drying process will occupy some days. No other preparation is needed; and they may be loosely packed in paper bags, or, what is preferable, spread out

Everything being now in readiness, you take each specimen separately out of the sea-water, where it must always remain until wanted ; and, taking care it is perfectly clean, place it in the dish of fresh water before alluded to. Having selected a piece of paper, cut to the dimensions corresponding with the size of the plant, taking care to leave space enough for a clear margin of three quarters of an inch from the edge of the paper, you place it in the palm of your left hand, and at once insert it under the plant floating in the water ; then, with the fingers of the other hand, assisted by the points of the closed forceps before mentioned, you proceed to separate the branches, and neatly arrange the specimen, endeavouring to preserve its natural appearance, and character of growth, as much as possible. Although some persons use a probe, or pointer, as well as a camel's-hair pencil, to display their specimens, I find the forceps alone sufficient. It often happens that there is occasion to remove one or more superfluous

between sheets of Bentall's botanical drying-paper, and not submitted to any pressure, observing to make a memorandum of the exact locality where they were gathered.

It is very much better, when drying Algæ in this rough manner, not to wash them in fresh water, because the salt they contain serves to keep them in a pliable state, and causes them to imbibe water more readily on re-immersion.

branches, which, if left, would only tend greatly to diminish the beauty of the specimen, and give it a very unsightly appearance when dried. Here the forceps serve in the place of a pair of scissors; and you can with them at once disengage all superfluous portions. I would here recommend the young algologist not to be sparing in pruning his plants, whenever it is required, as they always present a much neater appearance when laid out moderately thin. Besides, the fruit will be better shown, and the species more easily recognized.

The specimen being now properly arranged to your satisfaction, and kept as nearly as possible in the centre of the paper, it is to be gently raised from the water while still remaining on your hand, which must be inclined a little on one side for a short time, to allow the superabundant water to run off. It is then to be placed on one of the folded sheets of the coarser kind of drying-paper, where it must remain whilst you proceed to operate in a similar manner with other specimens, until there is no further space left on the blotting-paper. You now take a piece of thin calico, cambric muslin, or some other similar material, and place it carefully and very smoothly over the plants, and over that another folded sheet of drying-paper, upon which other plants are to be laid, in the manner before described. A reasonable number of layers of plants, say from eight to ten, being prepared, with pieces of calico and sheets of drying-paper interspersed between each layer, the whole is to be removed to one of the square pieces of deal; whilst another piece is to be placed on the top, and immediately subjected to a certain amount of pressure. When there are but few specimens, and these of the finer varieties, one or two books, or a flat stone, will be quite sufficient for this purpose; but, should there be a large quantity, and these happen to be of a coarser description, a screw-press is necessary, the proper amount of force required being readily ascertained by a little practice. Great caution, however, must be observed, that the pressure is not too severe, as the specimens would not only be liable to be injured, but the impression of the threads of the calico would be visible, and give them an unsightly and unnatural appearance. As a guide to the young algologist, and to prevent such a result, it should be remembered that it is best to commence at first, when the plants are wet, with a very slight degree of pressure; and to increase it gradually as they become nearly dry, when any amount of pressure may be used to advantage.

In about two or three hours' time, the wet layers of drying-paper should be removed, which must be done very carefully; but on no account must the pieces of calico be disturbed until the specimens are dry, and fresh folded sheets of the finer sort of drying-paper substituted in their place. Additional layers are to be employed until the plants are thoroughly dry, which, after the second change of papers, need not be renewed oftener than once in every twelve hours. When the plants are supposed to be dry, which, if properly attended to, generally occurs, except in some of the coarser species, in about three or four days, the pieces of calico are to be removed; this must be done very carefully, as it sometimes happens that portions of the plants will be found adhering to them.* After this is done, I usually complete the process by placing the specimens once more between layers of the drying-paper, and immediately submit them to a heavy pressure, for several days. Plants so treated never become damp or mouldy.

Before removing them to the herbarium, the name of the plant, and the place of its growth (or wherever it may have been found), as well as the day of the month and year, should always be written, in pencil, on the back of the specimen, which will be a guide to the collector, and enable him, when in want of such species at any future time, to know where to find them, although it occasionally happens that some plants are somewhat

---

* Should this occur, and it be found difficult to remove them from the calico, it is better to allow them to dry thoroughly, trusting to after removal, than to tear them away in a half-dried state, which would probably destroy the specimen.

uncertain as to the time of their appearance, and may be found earlier or later in some seasons than in others.

The young beginner, as well as the more experienced collector, will sometimes stumble upon a plant he has never seen before, and consequently will not be able to name it; and he may also be unprovided with a microscope sufficiently powerful to examine it, in order to establish its identity.

In this case, it is desirable that he should forward the specimen, with as little delay as possible, to some competent and scientific algologist, for a more elaborate investigation, as the structure of most plants by which their genera and species are determined, are far more easily made out whilst they are fresh, and recently gathered, than after being dried.* It is therefore advisable that any plant of doubtful character should, as soon as practicable after being removed from its place of growth, or wherever it may have been found, be forthwith placed between two pieces of rag, and afterwards

---

* It is often very difficult to examine the structure of the finer species, except immediately after being fresh gathered; but, should circumstances render this impracticable, in consequence of the plants having been already dried, a small fragment of the specimen must be moistened with water; and if the sections do not open well, by adding a drop of muriatic acid you will find they will expand very freely.

between two layers of tin-foil, or, what is still better, a portion of the thin leaves of gutta percha; then placed in an envelope, and despatched by post, without any unnecessary loss of time.*

Though the method I have just described may be generally adopted with the larger portion of Algæ, there are some particular species that require a different mode of treatment in laying out for the purpose of being dried, which I will now proceed to describe.

Most of the Melanospermeæ, or olivaceous series, give out a large quantity of mucus, after being immersed in fresh water, which renders it difficult to preserve good specimens in the ordinary way, from their adhering tenaciously both to the cloths and the paper. Amongst these, I may notice all the different species of Fucoideæ, Chorda Filum, Cystoseira, Halydrys siliquosa, Pycnophycus tuberculatus, Chondrus crispus, and others of similar texture. All the plants thus enumerated require to be treated in a particular manner, and differently from others.

Instead of attempting to display them immediately after being gathered, it is best to spread them out on

* In this manner, I am frequently in the habit of transmitting recent specimens to Professor Harvey, of Trinity College, Dublin; Mrs. Griffiths, of Torquay; and other correspondents: and I find they generally arrive in excellent condition.

coarse towels for a day or two, and allow them to wither
a little, when they will become more manageable, by
being less rigid. After this, they are to be put into a
deep dish, and a quantity of boiling water poured over
them, and allowed to remain for half an hour, or a little
more. This will deprive them of a great deal of mucus, when the plants are to be separately washed in
cold fresh water, and neatly spread between several
folds of calico or common towels, and thus allowed to
dry, exposing them to the air occasionally, to prevent
their becoming mouldy. When nearly dry, and whilst
they are still flexible, remove them to a vessel containing fresh water, where they may remain about half an
hour ; and they are then to be managed precisely in the
same manner as before described, until they are thoroughly dry.

As it will be found that several of the species just
referred to, and, indeed, many others, after being dried,
will not adhere to paper, the manner of fixing them remains to be pointed out ; and I particularly recommend
that this should be done with as little delay as possible,
as they may be irreparably injured, if they are not properly secured before being removed to the herbarium
or elsewhere.

With the finer sorts, I find a little skimmed milk is
quite sufficient to fix them firmly to the paper, taking

care, however, that every particle of cream or butter is first removed.

The milk is to be applied with a tolerably-sized, flat, camel's-hair pencil or varnish-brush, all over the paper and the specimen, as softly as possible; after which pieces of muslin or calico are to be spread over it, and, lastly, layers of drying-paper. It is then to be subjected to gentle pressure for twenty-four hours, when the muslin may be removed, and the plant, for the last time, submitted to considerable pressure for a few days.

The more robust species, such as the Fucoideæ, Cystoseiræ, &c., require a different application to fasten them to the paper. When this is necessary, I invariably use a solution of isinglass,* which should be kept in a glass-stoppered bottle; and, when wanted for use, the bottle is to be placed in any convenient vessel containing hot water, to render it fluid, taking care to remove the stopper beforehand. This solution is to be applied also,

---

* Take a quarter of an ounce of isinglass, and an ounce and a half of water, and put the same into a wide-mouthed bottle; place it in a water-bath, and, when thoroughly dissolved, add one ounce of rectified spirits of wine, heated nearly to the same temperature; stir the whole well together; and, when cold, keep it well stopped till it is required for use. This preparation I have used for many years, and have found it far preferable to gum-water, which, after a time, is apt to crack, and peel off, besides leaving a glossy mark wherever it is applied.

with a camel's-hair brush, to the under sides of the
stems and branches of the plant; after which it is to
be removed to the press; and, when dry, it will be
found to adhere to the paper.

CHAPTER III.

THOSE who aspire to extend their knowledge of marine botany beyond the mere arrangement of the easily recognized species, will require the aid of the microscope.

For identifying many species of Calithamnia, Ceramieæ, Ectocarpeæ, Polysiphoniæ, and others, and for most other purposes, simple microscopes will generally be found sufficient.

When collecting on the sea-shore, the algologist frequently meets with plants, such as Ceramiæ, Polysiphoniæ, &c., assuming such doubtful forms, that it is utterly impossible to decide to what species they properly belong, without the assistance of a glass of considerable magnifying power.

At such times, it is very desirable to have ready means for examination ; otherwise, one may be collecting and filling his bag with useless trash, under the impression of their being prizes ; or, on the other hand, throw away, probably, some choice or rare specimens, under the impression that they are common.

The fructification, too, in many of the minutely-branched and filamentous Algæ, can only be detected under the microscope,—an object of great importance to the scientific algologist, inasmuch as all plants in fruit are of greatly enhanced value compared with those that are barren.

For use at the sea-side, and for all ordinary purposes, the most convenient instrument is the *Stanhope lens.* After moistening the flat side, a fragment of the plant to be examined is to be placed on it, and then viewed by looking towards the light, with the convex side of the lens close to the eye.

It requires but very little practice to enable any person to understand its use ; it can be purchased at any optician's, the price varying from 2s. 6d. to 10s. ; may be worn suspended by a ribbon, like an eye-glass ; and requires no adjustment.

By its instrumentality, the collector will be able to observe beauties to which he might otherwise have been a stranger ; whilst his knowledge of the plants he collects will be greatly extended.

The perfect comprehension, however, of the appearances and peculiarities which characterize the nice distinctions botanists have assigned to genera and species, should be the aim of every one who aspires to be something more than a mere collector.

For such purposes, the Stanhope lens is not sufficient.

The algologist must now call in the aid of *doublets*, or of the compound microscope.

In examining plants with the compound microscope, if it be not intended to preserve the specimen, it is suf-ficient to place it on a slip of glass, with a drop of water, and cover it with another very thin piece of glass or mica, when it is ready for examination. When it is intended to preserve the specimen for further observation, or as a type of the species, the fragment must be mounted in a more permanent manner.

The plan most generally used by English microsco-pists, is to make cells on glass slides, by drawing lines with a camel's-hair brush dipped in varnish, or japanner's gold-size, so as to enclose circular or quadrangular spaces. When the varnish is dry, the cell is filled with some preservative fluid, in which the object is placed, and then covered with a thin piece of glass, of suitable size and figure. The edges of the *glass cover* are coated with varnish, to secure it to the slide, &c., and to prevent the preservative fluid from evaporating.

The glass slides and covers are sold at a moderate price by the opticians, assorted in boxes. The former are most commonly three inches long by one inch in width. The algologist should write on the end of the glass slide, with a writing-diamond, the genus and species, with such other particulars as date, place where found, &c., &c.

The preserving fluid should possess the following qualities :—It should be clear and colourless, not apt to become mouldy, or to generate minute Algæ ; it should have no action on the specimen, by altering its colour or texture, but should have the property of preserving the natural colour and appearance of the plant for a number of years.

The fluids recommended by the most practical microscopists, are *Goadby's preparations ;* the composition for the first being " One ounce of bay-salt, half an ounce of alum, one grain of corrosive sublimate, and one pint of boiling water.*   To be well stirred, and when cold, carefully filtered through blotting-paper."

" The second fluid is prepared by pouring one pint and three quarters of boiling water on eight ounces of bay-salt, and one grain of corrosive sublimate ; when cold, to be filtered through blotting-paper."

Mr. Ralfs, in his beautiful and elaborate work on the Desmidieæ, has given the composition of Mr. Thwaites' fluid as follows :—" Sixteen parts distilled water, one part rectified spirits of wine, and a few drops of creosote, mixed well together with a small quantity of prepared chalk, and then filtered.   This is to be afterwards

---

* Distilled water, which can be purchased at any respectable druggist's, should be used.

mixed with an equal quantity of distilled water saturated with camphor, and strained through fine linen."

Mr. Ralfs recommends bay-salt and alum, of each one grain, dissolved in one ounce of distilled water.

One ounce of acetate of alumina dissolved in four ounces of distilled water, is said to be a good preservative of delicate colours.

Canada balsam, for some purposes, has been frequently used; but, though this is a most excellent medium for many plants, it unfortunately renders the delicate tissues of some Algæ so extremely transparent, as to make them invisible under the microscope. Yet it has been found to answer extremely well for the brown marine Algæ (Melanospermeæ), as well as the red (Rhodospermeæ).

For the green series (Chlorospermeæ), and all the fresh-water Algæ, it is almost entirely inapplicable. With any plants having white filaments, or spines, whether among the red or green Algæ, there is the same objection.

I have now in my possession several specimens of Calithamnia, Griffithsiæ, Polysiphoniæ, &c., mounted, more than four years since, in Canada balsam, in which the colour is most beautifully preserved. The method to be pursued in mounting Algæ in any of the watery fluids, differs scarcely anything from the usual plan. It is as follows :—I use slips of glass (selecting good, flat, and clean crown-glass, free from specks) for the slides; and for the toppings, thin glass covers. The cells are thus prepared :—A card, having marked on it, in strong black lines, the outline of the intended cell, is to be laid under the glass slide, to serve as a guide in drawing the lines on the glass, with a ca-mel's-hair brush dipped in black Japan varnish. The brush should be very full of colour, to insure sufficient depth in the walls of the cell. The outline being drawn, the glass slide is to be laid aside, in a horizontal posi-tion, for a few hours, until the varnish is dry enough to permit the slide being placed edgeways without risk of its running. It is desirable to have a good many slides ready prepared, with cells of different sizes, and diffe-rent depth of walls.

When it is intended to mount a specimen, the best and most interesting portion of the plant is selected, and arranged carefully in the cell, which can only be properly done under a good single magnifier; the ramuli being accurately separated and spread, and not overlapping one another. Transverse and longitudinal sections of the plant, and a portion showing the fruit, &c., may also be placed in the same cell, properly dis-played.

A small quantity of the preserving fluid is then to be taken out of the bottle, by means of an ivory or glass bodkin, and the cell filled with it. The cell should be quite full, but not to overflowing.

The specimen should be again examined, to see that it has not been displaced in introducing the fluid. The glass cover is now to be laid on, with great care and steadiness, and adjusted perfectly square, to meet the walls of the cell. The superabundant fluid which will escape outside, should be absorbed and removed by a camel's-hair brush, observing that you do not thus take up too much liquid, or air will insinuate itself between the glasses; and this must be avoided. The glasses should be then laid aside for a short time, to allow all moisture to dry from the edges; after which a camel's-hair brush, charged with varnish, is to be drawn along the same, encroaching about half a line on the cover, and a line on the slide. This last process must be repeated two or three times, allowing an interval of four or five hours to elapse between each application, until a sufficient body of varnish is laid on, so that the edges of the cover may be firmly and properly cemented to the slide, and effectually prevent the evaporation of the fluid.

The mounted specimen must be left in a horizontal position about twenty-four hours, when it may be laid aside as finished: but as Japan varnish will take some time before it gets properly dry and hard, care must be taken that nothing rests against it until it is so. The edges of the varnished lines may be trimmed with a penknife, or some other light instrument, if thought necessary, after the varnish is perfectly dry.

To mount Algæ in Canada balsam, it is not necessary to prepare any cells : the object is merely to be laid on the glass slide, in as fresh a state as possible, to insure a proper degree of plumpness and fulness. Though it may be advisable to pass the specimen quickly through clean fresh water, to remove as much of the salt as possible, it must be borne in mind that some plants, such as the Griffithsiæ, will not bear fresh water for an instant. The specimen, after being displayed with great care, and as expeditiously as possible, is then to be left until all the external water shall have evaporated from its surface, as well as from the glass.

A bodkin or probe, or any other suitable instrument, is then to be dipped very gently (to avoid air-bubbles) into the Canada balsam, and then withdrawn rather more rapidly. A quantity of the balsam will adhere to the instrument; and by holding it vertically a drop will be formed at the lower end. The apex of the drop is then to be brought in contact with the specimen, as near the centre as possible. By degrees the drop will detach itself from the instrument, when the latter must be withdrawn. The drop should be large enough to cover the specimen completely after being spread out and properly arranged beforehand, in order to avoid the necessity of applying another drop, thereby involving the risk of air-bubbles. This operation having been satisfactorily achieved, the thin glass cover is to be laid on, placing it on the apex of the drop; and then on the cover lay a piece of stout sheet-lead, of nearly the same dimensions, which will gradually press out the balsam between the glasses. These are now to be laid aside, in a horizontal position, for a week, when the lead weight may be removed. The name may then be written on the glass; and it is not advisable to place the slide vertically, or to trim the edges, for the space of a month, as the balsam dries very slowly.

This plan is recommended to those who may be desirous to investigate marine Algæ microscopically.

It is, indeed, at all times very desirable to possess even but a fragment of a duly-authenticated specimen, with which to compare one of our own collection. The collector will find no more certain plan for effectual comparison, than to take a glass slide, with its mounted specimen, and place beside it a fragment of the plant under inspection, and examine both at the same time, with a good single magnifier.

Comparing recent with dried specimens will not often be very satisfactory, because many plants alter much in their appearance in drying; whereas the mounted specimen, if properly prepared, will generally preserve its natural form and similitude; and then, on comparing it with fresh specimens, its peculiar character will be observed.

In mounting with Canada balsam, the great art appears to consist in laying on the balsam at the right moment, that is, when the plant and edges are not too moist; for, when this happens, a sort of dew will be formed on the balsam : and, on the other hand, if too dry, the cells and fine ramuli of the specimen will have become considerably shrunk.

It need scarcely be remarked that, in whatever way the mounting may be performed, the specimens should be previously cleaned of parasites, whether other species of Algæ, Diatomaceæ, or zoophytes. Though it has been before stated that Canada balsam is not suited for the Chlorospermeæ, yet for some of the larger and coarser weeds it answers exceedingly well, and shows the cells very distinctly.

If it is required to mount specimens of plants already dried, in Canada balsam, they must be immersed for a few minutes in sea-water, then rinsed in fresh water, and afterwards treated in the same manner as recent specimens.

A LIST

OF THE

BRITISH MARINE ALGÆ,

ARRANGED SYSTEMATICALLY,

AFTER

PROFESSOR HARVEY'S MANUAL.

SUB-CLASS I.

MELANOSPERMEÆ, or FUCALES.

Olive Sea-weeds.

Order I. FUCACEÆ.

Olive-coloured, inarticulate sea-weeds, whose spores are contained in spherical cavities of the frond.

I. Sargassum.

Branches bearing ribbed leaves.

1. Sargassum vulgare.
2.   ,,   bacciferum.

II. Halidrys.

Frond linear, pinnate, leafless.

1. Halidrys siliquosa.

III. Cystoseira.

Root scutate.   Frond much branched, bushy.

1. Cystoseira ericoides.
2.   ,,   granulata.
3.   ,,   barbata.
4.   ,,   fœniculacea.
5.   ,,   fibrosa.

IV. Pycnophycus.

Root branching.   Frond cylindrical.

1. Pycnophycus tuberculatus.

V. Fucus.

Root scutate. Frond dichotomous.

1. Fucus vesiculosus.
2.　　,,　　ceranoides.
3.　　,,　　serratus.
4.　　,,　　nodosus.
5.　　,,　　Mackaii.
6.　　,,　　canaliculatus.

VI. Himanthalia.

Root scutate. Frond cup-shaped.

1. Himanthalia lorea.

————

Order II. SPOROCHNACEÆ.

Olive-coloured, inarticulate sea-weeds, whose spores are attached to externally-jointed filaments.

VII. Desmarestia.

Frond solid, distichous, filiform, or flat.

    1. Desmarestia ligulata.
    2.   ,,   aculeata.
    3.   ,,   viridis.

VIII. Arthrocladia.

Frond filiform, nodose, traversed by a jointed tube.

    1. Arthrocladia villosa.

IX. Sporochnus.

Frond filiform. Receptacles lateral, stalked.

    1. Sporochnus pedunculatus.

X. Carpomitra.

Frond filiform or flat, and mid-ribbed. Root a shapeless, woolly tuber. Receptacles terminal, mitriform.

    1. Carpomitra Cabreræ.

Order III. LAMINARIACEÆ.

Olive-coloured, inarticulate sea-weeds, whose spores are superficial, either forming cloud-like patches, or covering the whole surface of the frond.

XI. ALARIA.

Frond stipitate. Stipes ending in a midribbed leaf. Root fibrous.

    1. Alaria esculenta.

XII. LAMINARIA.

Frond stipitate. Stipes ending in a ribless leaf.

    1. Laminaria digitata.
    2.   ,,   bulbosa.
    3.   ,,   saccharina.
    4.   ,,   Phyllitis.
    5.   ,,   fascia.

XIII. CHORDA.

Frond leafless, cylindrical, hollow, the cavity interrupted by transverse partitions. Root scutate.

1. Chorda Filum.
2.   „   Lomentaria.

Order IV. DICTYOTACEÆ.

Olive-coloured, inarticulate sea-weeds, whose spores are superficial, disposed in definite spots or lines (sori).

XIV. Cutleria.

Frond ribless, irregularly cleft. Sori dot-like, scattered. Root clothed with woolly fibres.

    1   Cutleria multifida.

XV. Haliseris.

Frond flat, linear, membranaceous, with a mid-rib. Root a mass of woolly filaments.

    1. Haliseris polypodioides.

XVI. Padina.

Frond wedge-shaped at base, fan-shaped, concentrically striate. Root coated with woolly fibres.

    1. Padina Pavonia.

XVII. Zonaria.

Frond ribless, flat, fan-shaped. Root coated with woolly fibres.

     1. Zonaria parvula.

XVIII. Taonia.

Frond ribless, irregularly cleft, somewhat fan-shaped, marked with concentric lines. Sori on both surfaces of the frond. Root coated with woolly fibres.

     1. Taonia atomaria.

XIX. Dictyota.

Frond ribless, dichotomous. Spores scattered irregularly, or clustered. Root coated with woolly fibres.

     1. Dictyota dichotoma.

XX. Stilophora.

Frond filiform, irregularly branched. Spores concealed among moniliform threads. Root a small, naked disk.

     1. Stilophora rhizodes.
     2.   ,,   Lyngbyei.

XXI. Dictyosiphon.

Frond filiform, tubular, branched. Spores irregularly scattered, solitary, or in dot-like sori. Root a small, naked disk.

1. Dictyosiphon fœniculaceus.

XXII. Striaria.

Frond filiform, tubular. Spores in dot-like sori, ranged in transverse lines. Root naked and scutate.

1. Striaria attenuata.

XXIII. Punctaria.

Frond flat, leaf-like. Spores scattered over the whole frond, in minute, distinct dots.

1. Punctaria latifolia.
2. ,, plantaginea.
3. ,, tenuissima.

XXIV. Asperococcus.

Frond membranaceous, tubular, or compressed. Spores in dot-like sori. Root minutely scutate, naked.

1. Asperococcus compressus.
2.   „   Turneri,
3,   „   echinatus.

XXV. Litosiphon.

Frond cartilaginous, unbranched, cylindrical, composed of several rows of cells. Spores scattered. Growing parasitically on Chorda Filum.

1. Litosiphon pusillus.
2.   „   Laminariæ.

Order V. CHORDARIACEÆ.

Olive-coloured sea-weeds, with a gelatinous or cartilaginous frond, composed of vertical and horizontal filaments interlaced together.

XXVI. Chordaria.

Frond filiform, much branched, cartilaginous.

1. Chordaria flagelliformis.
2.   „   divaricata.

E

XXVII. Mesogloia.

Frond filiform, much branched, gelatinous.

　1. Mesogloia vermicularis.
　2.　　,,　　Griffithsiana.
　3.　　,,　　virescens.

XXVIII. Leathesia.

Frond tuber-shaped. Spores attached to the coloured tips of the filament.

　1. Leathesia tuberiformis (Corynephora marina).
　2.　　,,　　Berkeleyi.

XXIX. Ralfsia.

Frond crustaceous. Fructification depressed warts scattered over the upper surface, containing obovate spores.

　1. Ralfsia verrucosa (R. deusta).

XXX. Elachistea.

Frond parasitical, consisting of a dense tuft of olivaceous filaments, arising from a tubercular base, composed of vertical fibres.

1. Elachistea fucicola.
2.   ,,   flaccida.
3.   ,,   curta.
4.   ,,   pulvinata (attenuata).
5.   ,,   stellulata.
6.   ,,   scutulata.
7.   ,,   velutina.

XXXI. MYRIONEMA.

Minute parasites, consisting of a mass of short, erect, simple jointed filaments, spreading, in patches, on the surface of other Algæ.

1. Myrionema strangulans.
2.   ,,   Leclancherii.
3.   ,,   punctiforme.
4.   ,,   clavatum.

Order VI. ECTOCARPACEÆ.

Olive-coloured, articulated sea-weeds, filiform. Spores (generally) external, attached to the jointed ramuli.

E 2

XXXII. CLADOSTEPHUS.

Frond inarticulate. Ramuli whorled. Spores borne by accessory ramuli.

     1. Cladostephus verticillatus.
     2.   ,,   spongiosus.

XXXIII. SPHACELARIA.

Frond flaccid. Ramuli distichously branched, mostly pinnated. Oval spores borne on the ramuli.

     1. Sphacelaria filicina.
     2.   ,,   Sertularia.
     3.   ,,   Scoparia.
     4.   ,,   plumosa.
     5.   ,,   cirrhosa.
     6.   ,,   fusca (olivacea).
     7.   ,,   radicans.
     8.   ,,   racemosa.

XXXIV. ECTOCARPUS.

Frond branching. Ramuli scattered.

     1. Ectocarpus siliculosus.
     2.   ,,   amphibius.
     3.   ,,   fenestratus.

4. Ectocarpus fasciculatus.
5.   ,,   Hincksii.
6.   ,,   tomentosus.
7.   ,,   crinitus.
8.   ,,   pusillus.
9.   ,,   distortus.
10.   ,,   Landsburgii.
11.   ,,   littoralis.
12.   ,,   longifructus.
13.   ,,   granulosus.
14.   ,,   sphærophorus.
15.   ,,   brachiatus.
16.   ,,   Mertensii.

XXXV. MYRIOTRICHIA.

Frond unbranched. Ramuli whorled. Elliptical spores containing a dark-coloured granular mass.

1. Myriotrichia clavæformis.
2.   ,,   filiformis.

SUBCLASS II.

RHODOSPERMEÆ, or CERAMIALES.

Order VII. RHODOMELACEÆ.

Red or purple, rarely brown-red, sea-weeds. Fructification of two kinds :—1. Spores contained either in external or immersed conceptacles, or densely aggregated together, and dispersed in masses through the substance of the frond ; 2. Spores (called tetraspores) immersed in distorted ramuli, or contained in proper receptacles.

XXXVI. ODONTHALIA.

Frond plano-convex, distichous, alternately toothed at the margin. Fructification twofold.

    1. Odonthalia dentata.

XXXVII. RHODOMELA.

Frond filiform, much branched, and coated with minute irregular cells. Fructification twofold on distinct plants.

    1. Rhodomela lycopodioides.
    2.   ,,   subfusca.

XXXVIII. Bostrychia.

Frond dull purple, filiform, much branched, dotted. Apices strongly involute. Fructification twofold.

1. Bostrychia scorpioides.

XXXIX. Rytiphlæa.

Frond filiform or compressed. Branches transversely striate. Fructification twofold on distinct plants.

1. Rytiphlæa pinastroides.
2.   ,,   complanata.
3.   ,,   thuyoides.
4.   ,,   fruticulosa.

XL. Polysiphonia.

Frond filamentous, partially or generally articulate. Articulations of the ramuli two or many-tubed. Tetraspores on distorted ramuli.

1. Polysiphonia urceolata.
      ,,   var. *patens.*
2.   ,,   formosa.
3.   ,,   stricta.
4.   ,,   pulvinata.
5.   ,,   fibrata.
6.   ,,   spinulosa.

7. Polysiphonia Richardsoni.
8.   ,,   Griffithsiana.
9.   ,,   elongella.
10.   ,,   elongata.
11.   ,,   Grevillii.
12.   ,,   violacea.
13.   ,,   Carmichaeliana
14.   ,,   fibrillosa.
15.   ,,   Brodiæi.
16.   ,,   variegata.
17.   ,,   obscura.
18.   ,,   simulans.
19.   ,,   nigrescens.
20.   ,,   affinis.
21.   ,,   subulifera.
22.   ,,   atro-rubescens.
23.   ,,   furcellata.
24.   ,,   fastigiata.
25.   ,,   parasitica.
26.   ,,   byssoides.

XLI. Dasya.

Frond filamentous. The stem and branches mostly opaque. Articulations of the ramuli single-tubed. Tetraspores in lanceolate, pod-like receptacles.

1. Dasya coccinea.
2.   ,,   ocellata.
3.   ,,   arbuscula.
4.   ,,   venusta.

Order VIII. LAURENCIACEÆ.

Rose-red or purple sea-weeds, with a cylindrical or compressed branching frond. Fructification double :—
1. Conceptacles containing a tuft of pear-shaped spores ;
2. Tetraspores immersed in the branches and ramuli, scattered, without order, through the surface-cells.

XLII. BONNEMAISONIA.

Frond filiform, extremely branched. The branches margined with distichous, awl-shaped, alternate cilia. Tetraspores unknown.

1. Bonnemaisonia asparagoides.

XLIII. LAURENCIA.

Frond cylindrical or compressed, purplish, yellowish, or reddish. Structure cellular, solid. Fructification of two kinds on distinct individuals.

1. Laurencia pinnatifida.
2.   ,,   cæspitosa (hybrida).
3.   ,,   obtusa.
4.   ,,   dasyphylla.
5.   ,,   tenuissima.

XLIV. Chrysimenia.

Frond tubular, continuous, not constricted or jointed. Fructification of two kinds.

1. Chrysimenia clavellosa.
2.   ,,   Orcadensis.

XLV. Chylocladia.

Frond (at least, the branches) tubular, constricted at regular intervals, and divided, by internal diaphragms, into chambers. Fructification double.

1. Chylocladia ovalis.
2.   ,,   kaliformis.
3.   ,,   reflexa.
4.   ,,   parvula.
5.   ,,   articulata,

Order IX. CORALLINACEÆ.

Rigid, articulated, or crustaceous, mostly calcareous sea-weeds, purple when recent, fading, on exposure, to white, composed of cells, in which carbonate of lime is deposited in an organized form. Tetraspores tufted, contained in spherical conceptacles.

XLVI. CORALLINA.

Frond filiform, articulated, branched, mostly pinnate, coated with a calcareous deposit.

1. Corallina officinalis.
2.   ,,   elongata.
3.   ,,   squamata.

XLVII. JANIA.

Frond filiform, articulated, dichotomous, branched, coated with a calcareous deposit.

1. Jania rubens.
2.   ,,   corniculata.

XLVIII. MELOBESIA.

Frond attached or free, opaque, covered with a calcareous deposit. Tetraspores scattered over the surface of the frond.

1. Melobesia polymorpha.
2.   ,,   calcarea.
3.   ,,   fasciculata.
4.   ,,   agariciformis.
5.   ,,   lichenoides.
6.   ,,   membranacea.
7.   ,,   farinosa.
8.   ,,   verrucata.
9.   ,,   pustulata.

XLIX. Hildebrandtia.

Frond cartilaginous (not stony), composed of very slender, closely-packed, vertical filaments.

1. Hildebrandtia rubra.

L. Lithocystis.

Plant calcareous, consisting of a single plane of cellules, which are disposed in radiating, dichotomous series, forming an appressed, flabelliform frond.

1. Lithocystis Allmanni.

Order X. DELESSERIACEÆ.

Rosy or purplish red, or blood-red, sea-weeds, with a leafy or rarely filiform, areolated, inarticulate frond. Leaves delicately membranaceous. Fructification double.

LI. DELESSERIA.

Frond rose-red, flat, membranaceous, with a percurrent mid-rib. Fructification of two kinds on distinct individuals.

1. Delesseria sanguinea.
2.   ,,   sinuosa.
3.   ,,   alata.
4.   ,,   angustissima.
5.   ,,   Hypoglossum.
6.   ,,   ruscifolia.

LII. NITOPHYLLUM.

Frond membranaceous, reticulated, rose-red, veinless, without mid-rib. Fructification double.

1. Nitophyllum punctatum.
2.   ,,   Hilliæ.

3. Nitophyllum Bonnemaisonii.
4.   ,,   Gmelini.
5.   ,,   laceratum.
6.   ,,   versicolor.

LIII. Plocamium.

Frond pinky-red, linear, compressed or flat, ribless, distichously much branched. The ramuli alternate, acute. Fructification of two kinds.

1. Plocamium coccineum.

—————

Order XI. RHODYMENIACEÆ.

Purplish or blood-red sea-weeds, with an expanded or filiform, inarticulate frond. Fructification double :—1. Conceptacles external, or half immersed, containing, beneath a thick pericarp, a mass of spores ; 2. Tetraspores either dispersed through the whole frond, or collected in indefinite, cloudy patches.

LIV. Stenogramme.

Frond rose-red, nerveless, laciniate. Fructification linear, convex, longitudinal (nerve-like) conceptacles, containing a dense mass of minute spores.

    1. Stenogramme interrupta.

LV. Rhodymenia.

Frond flat, membranaceous, ribless, veinless, cellular. Fructification :—1. Convex tubercles, containing a mass of minute spores ; 2. Tetraspores imbedded among the cells of the surface, scattered or forming cloudy patches.

    1. Rhodymenia bifida.
    2.   ,,   laciniata.
    3.   ,,   Palmetta.
    4.   ,,   cristata.
    5.   ,,   ciliata.
    6.   ,,   jubata.
    7.   ,,   palmata.

LVI. Sphærococcus.

Frond cartilaginous, compressed, two-edged, distichously branched. Fructification spherical tubercles, containing a mass of minute spores.

    1. Sphærococcus coronopifolius.

LVII. Gracilaria.

Frond filiform (rarely flat), cellular. Fructification :—
1. Convex tubercles, containing a mass of minute spores;
2. Tetraspores imbedded in the cells of the surface.

1. Gracilaria multipartita.
2.   ,,   compressa.
3.   ,,   confervoides.
4.   ,,   erecta.

LVIII. Hypnea.

Frond filiform, cartilaginous, much branched, cellular.
Fructification:—1. Spherical tubercles, immersed in the
ramuli; 2. Tetraspores imbedded in the surface-cells.

1. Hypnea purpurascens.

Order XII. CRYPTONEMIACEÆ.

Purplish or rose-red sea-weeds, with a filiform or
(rarely) expanded, gelatinous or cartilaginous frond.
Fructification :—1. Globose masses of spores immersed
in the frond, or in swellings of the branches; 2. Tetraspores variously dispersed.

LIX. Grateloupia.

Frond pinnated, flat, narrow, cartilaginous, of very dense structure. Favellidia immersed in the branches. Tetraspores scattered.

1. Grateloupia filicina.

LX. Gelidium.

Frond linear, compressed, horny, of a very dense structure. Fructification :—1. Favellidia immersed in swollen ramuli ; 2. Tetraspores immersed in the ramuli.

1. Gelidium corneum.
    β.   ,,   sesquipedale.
    γ.   ,,   pinnatum.
    δ.   ,,   uniforme.
    ε.   ,,   capillaceum.
    ζ.   ,,   latifolium.
    η.   ,,   confertum.
    θ.   ,,   aculeatum.
    ι.   ,,   abnorme.
    κ.   ,,   pulchellum.
    λ.   ,,   claviferum.
    μ.   ,,   clavatum.
    ν.   ,,   crinale.
2.   ,,   cartilagineum.

F

LXI. Gigartina.

Frond cartilaginous, irregularly branched, purple or very dark red, filiform, compressed or flat. Favellidia in external tubercles. Tetraspores contained in dense, immersed sori.

1. Gigartina pistillata:
2.   ,,   acicularis.
3.   ,,   Teedii.
4.   ,,   mamillosus.

LXII. Chondrus.

Frond cartilaginous, nerveless, compressed or flat, dichotomously cleft, of very dense structure. Tetraspores collected into sori, immersed in the frond, and scattered over its segments.

1. Chondrus crispus.
2.   ,,   Norvegicus.

LXIII. Phyllophora.

Frond stipitate, rigid, membranaceous, nerveless, of very dense structure. Tetraspores in superficial sori, or in proper leaflets.

1. Phyllophora rubens.
2.   ,,   membranifolius.
3.   ,,   Brodiæi.
4.   ,,   Palmettoides.

LXIV. Peyssonelia.

Frond brownish red, depressed, rooting by the under surface. Tetraspores contained in superficial warts over the upper surface of the frond.

1. Peyssonelia Dubyi.

LXV. Gymnogongrus.

Frond cylindrical or compressed, horny, of very dense structure, much branched. Tetraspores in superficial warts.

1. Gymnogongrus Griffithsiæ.
2.   ,,   plicatus.

LXVI. Polyides.

Root an expanded disk. Frond cylindrical, dichotomous, cartilaginous, solid. Fructification:—1. Oblong, irregular, spongy warts; 2. Tetraspores cruciate, immersed in the branches.

1. Polyides rotundus.

LXVII. Furcellaria.

Root branching. Frond cylindrical, dichotomous, cartilaginous. Favellæ immersed in the pod-like swollen extremities of the branches. Tetraspores similarly immersed.

1. Furcellaria fastigiata.

LXVIII. Dumontia.

Frond cylindrical, tubular. Fructification:—1. Clusters of obovate spores, attached to the inner surface of the membrane of the frond; 2. Roundish tetraspores among the surface-cells.

1. Dumontia filiformis.

LXIX. Halymenia.

Frond compressed or flat, gelatinous, membranaceous. Fructification spherical masses of spores immersed in the frond.

1. Halyménia ligulata.
     ,,   var. *latifolia.*

LXX. Ginnania.

Frond cylindrical, tubular, distended, traversed by a fibrous axis. Fructification spherical masses immersed in the frond.

    1. Ginnania furcellata.

LXXI. Kalymenia.

Frond blood-red, ribless, expanded, leaf-like, thick, of dense structure. Fructification :—1. Spherical masses of spores semi-immersed in the frond ; 2. Scattered tetraspores.

    1. Kalymenia reniformis.
    2.   ,,   Dubyi.

LXXII. Iridæa.

Frond expanded, leaf-like, thick, of dense structure, dull red. Fructification :— 1. Spherical masses of spores immersed in the frond ; 2. Tetraspores scattered.

    1. Iridæa edulis.

LXXIII. Catenella.

Frond tubular, branched, constricted at intervals, dull purple. Fructification of two kinds.

1. Catenella Opuntia.

LXXIV. Cruoria.

Frond gelatinous, coriaceous, forming a skin on the surface of rocks, composed of vertical, articulated filaments.

1. Cruoria pellita.

LXXV. Naccaria.

Frond filiform, solid, rose-red. Ramuli composed of jointed, dichotomous, whorled filaments. Fructification, spores attached to the whorled filaments of the ramuli.

1. Naccaria Wigghii.

LXXVI. Gloiosiphonia.

Frond cylindrical, filiform, somewhat gelatinous, the walls composed of radiating filaments. Fructification, globules of red spores.

1. Gloiosiphonia capillaris. .

LXXVII. NEMALEON.

Frond filiform, gelatinous, cartilaginous, elastic, solid. Fructification, globular masses of spores (favellidia) attached to the filaments of the periphery.

1. Nemaleon multifidum.
2.   ,,   purpureum.

LXXVIII. DUDRESNAIA.

Frond cylindrical, gelatinous, elastic. Fructification : —1. Globular masses of spores attached to the filaments of the periphery ; 2. External tetraspores borne by the filaments of the periphery.

1. Dudresnaia coccinea.
2.   ,,   Hudsoni (*divaricata*).

LXXIX. CROUANIA.

Frond filiform, consisting of a jointed, single-tubed filament, whorled at the joints with minute multifid ramelli. Fructification, tetraspores attached to the base of the ramuli.

1. Crouania attenuata.

Order XIII. CERAMIACEÆ.

Rose-red or purple sea-weeds, with a filiform frond, consisting of an articulating, branching filament, composed of a single string of cells. Fructification:—1. Favellæ, berry-like receptacles, containing numerous angular spores ; 2. Tetraspores attached to the ramuli, or more or less immersed in the substance of the branches.

LXXX. PTILOTA.

Frond compressed, inarticulate, distichous. Fructification :—1. Roundish, clustered favellæ, surrounded by an involucre of short ramuli ; 2. Tetraspores attached to or immersed in the ultimate pinnules.

    1. Ptilota plumosa.
    2.   ,,   sericea.

LXXXI. MICROCLADIA.

Frond filiform, compressed, inarticulate, dichotomous, Fructification :—1. Roundish favellæ ; 2. Tetraspores immersed in the ramuli.

    1. Microcladia glandulosa.

LXXXII. Ceramium.

Frond filiform, one-tubed, articulated; the dissepiments coated with a stratum of coloured cellules, which sometimes extend over the surface of the articulation. Fructification :—1. Sessile, roundish favellæ, having a pellucid limbus, containing minute spores ; 2. Tetraspores either immersed in the ramuli, or more or less external.

1. Ceramium rubrum.
2.   ,,   botryocarpum.
3.   ,,   decurrens.
4.   ,,   Deslongchampii.
5.   ,,   diaphanum.
6.   ,,   gracillimum.
7.   ,,   strictum.
8.   ,,   nodosum.
9.   ,,   fastigiatum.
10.   ,,   flabelligerum.
11.   ,,   echionotum.
12.   ,,   acanthonotum.
13.   ,,   ciliatum.

LXXXIII. Spyridia.

Frond filiform, inarticulate, much branched, and clothed with minute articulated ramelli. Fructification:

1. Stalked, involucrate, lobed favellæ ; 2. Tetraspores attached to the ramuli.

　　　1. Spyridia filamentosa.

LXXXIV. Griffithsia.

Frond rose-red, filamentous. Filaments articulated throughout, dichotomous. Ramuli single-tubed, often whorled. Dissepiments hyaline. Fructification :—1. Roundish, gelatinous receptacles (favellæ), including minute granules ; 2. Tetraspores affixed to whorled ramuli.

　　　1. Griffithsia equisetifolia.
　　　2.　　,,　　simplicifilum.
　　　3.　　,,　　barbata.
　　　4.　　,,　　Devoniensis.
　　　5.　　,,　　corallina.
　　　6.　　,,　　secundiflora.
　　　7.　　,,　　setacea.

LXXXV. Wrangelia.

Frond rose-red, filamentous, jointed. Fructification : —1. Gelatinous receptacles (favellæ) ; 2. Tetraspores affixed to the ramuli, scattered.

　　　1. Wrangelia multifida.
　　　　　,,　　var. *pilifera*.

LXXXVI. Seirospora.

Frond rosy, filamentous. Stem articulated, the articulations traversed by jointed filaments. Tetraspores disposed in terminal, moniliform strings.

1. Seirospora Griffithsiana.

LXXXVII. Calithamnion.

Frond brownish red or rosy-red, filamentous. Branches and ramuli articulate, mostly pinnate. Fructification :—1. Roundish or lobed, berry-like receptacles on the main branches, containing spores ; 2. External tetraspores scattered along the ultimate branches, or borne on little stalks.

1. Calithamnion plumula.
2.   ,,   cruciatum.
3.   ,,   floccosum.
4.   ,,   Turneri.
5.   ,,   barbatum.
6.   ,,   Pluma.
7.   ,,   Arbuscula.
8.   ,,   Brodiæi.
9.   ,,   tetragonum.
10.   ,,   brachiatum.
11.   ,,   tetricum.

12. Calithamnion Hookeri.
13.   ,,   roseum.
14.   ,,   byssoideum.
15.   ,,   polyspermum.
16.   ,,   purpurascens.
17.   ,,   fasciculatum.
18.   ,,   Borreri.
19.   ,,   affine.
20.   ,,   tripinnatum.
21.   ,,   gracillimum.
22.   ,,   thuyoideum.
23.   ,,   corymbosum.
24.   ,,   spongiosum.
25.   ,,   pedicellatum.
26.   ,,   Rothii.
27.   ,,   floridulum.
28.   ,,   mesocarpum.
29.   ,,   sparsum.
30.   ,,   Daviesii.
31.   ,,   virgatulum.

CHLOROSPERMEÆ, or CONFERVALES.

Order XIV. SIPHONACEÆ.

Plants green, composed of continuous, tubular, or branched filaments (elongated, cylindrical cells, connected together into threads or filaments). Fructification :—1. Spores green or purple, formed within the cells, often, at maturity, vivacious, moving by means of vibratile cilia ; 2. External vesicles containing a dense, dark-coloured, granular mass, and finally separating from the frond.

LXXXVIII. Codium.

Filaments combined into a spongy frond. Fructification opaque vesicles attached to the filaments near the surface of the frond.

1. Codium Bursa.
2.   ,,   adhærens.
3.   ,,   amphibium.
4.   ,,   tomentosum.

LXXXIX. Bryopsis.

Frond filiform, cylindrical, glistening, branched. Branches pinnated, filled with a fine green, minutely granuliferous fluid.

1. Bryopsis plumosa.
2.   ,,   hypnoides.

XC. Vaucheria.

Frond aggregated, tubular, capillary, coloured by an internal, green, pulverulent mass. Fructification dark-green, homogeneous sporangia, attached to the frond.

1. Vaucheria submarina.
2.   ,,   marina.
3.   ,,   velutina.

Order XV. CONFERVACEÆ.

Green Algæ, composed of articulated filaments, simple or branched. Cells cylindrical.

XCI. Cladophora.

Filaments branched, composed of a single series of cells or articulations. Fruit aggregated granules or zoospores, contained in the articulations, having at some period a proper ciliary motion.

1. Cladophora Brownii.
2.   ,,   repens.
3.   ,,   pellucida.
4.   ,,   rectangularis.
5.   ,,   Macallana.
6.   ,,   Hutchinsiæ.
7.   ,,   diffusa.
8.   ,,   nuda.
9.   ,,   rupestris.
10.   ,,   lætevirens.
11.   ,,   flexuosa.
12.   ,,   gracilis.
13.   ,,   Rudolphiana.
14.   ,,   refracta.
15.   ,,   albida.
16.   ,,   lanosa.
17.   ,,   uncialis.
18.   ,,   arcta.
19.   ,,   glaucescens.
20.   ,,   falcata.

21. Cladophora flavescens.
22.   „   fracta.

XCII. Rhizoclonium.

Filaments jointed, decumbent. Branches short and
root-like. Fruit, granules contained in the cells.

1. Rhizoclonium riparia.

XCIII. Conferva.

Filaments unbranched, composed of a single series of
cells or articulations. Fruit aggregated granules or
zoospores contained in the articulations, having at some
period a proper ciliary motion.

1. Conferva arenicola.
2.   „   arenosa.
3.   „   littorea.
4.   „   Linum.
5.   „   sutoria.
6.   „   tortuosa.
7.   „   implexa.
8.   „   melagonium.
9.   „   ærea.
10.   „   collabens.
11.   „   bangioides.

12. Conferva Youngana.

13.   ,,   clandestina.

XCIV. Ochlochæte.

Frond disciform. Filaments radiating from a central point, consisting of a single series of cells, each cell produced above into a rigid, inarticulate bristle. Fructification unknown.

1. Ochlochæte hystrix.

Order XVI. ULVACEÆ.

Green; composed of small polygonal cells, forming membranous tubes.

XCV. Enteromorpha.

Frond tubular, of a green colour, and reticulated structure. Fructification, three or four roundish granules.

1. Enteromorpha Cornucopiæ.

2.   ,,   intestinalis.

3.   ,,   compressa.

4.   ,,   Linkiana.

G

5. Enteromorpha erecta.
6.   ,,   clathrata.
7.   ,,   ramulosa.
8.   ,,   Hopkirkii.
9.   ,,   percursa.

XCVI. ULVA.

Frond membranaceous, flat, of a green colour. Fructification minute granules.

1. Ulva latissima.
2. ,,   Lactuca.
3. ,,   Linza.

XCVII. PORPHYRA.

Frond flat, purple, exceedingly thin. Fructification: —1. Scattered sori of oval seeds; 2. Roundish granules, mostly covering the whole frond.

1. Porphyra laciniata.
2.   ,,   vulgaris.
3.   ,,   miniata.

XCVIII. BANGIA.

Frond filiform, tubular, mostly purple or pink. Spores purple or green.

1. Bangia fusco-purpurea.
2.   ,,   ciliaris.
3.   ,,   ceramicola.
4.   ,,   carnea.
5.   ,,   elegans.

Order XVII. OSCILLATORIACEÆ.

Green or blue; composed of continuous, tubular, simple, or rarely branching filaments.

XCIX. RIVULARIA.

Filaments radiating from a point, immersed in gelatinous, globose fronds.

1. Rivularia plicata.
2.   ,,   atra.
3.   ,,   applanata.
4.   ,,   nitida,

C. SCHIZOSIPHON.

Filaments sheathed, the sheath multifid.

1. Schizosiphon Warreniæ.

G 2

CI. SCHIZOTHRIX.

Filaments involved in a thick lamellar sheath, rigid, curled, thickened at the base.

    1. Schizothrix Creswellii.

CII. CALOTHRIX.

Filaments short, tufted, fixed at the base only.

    1. Calothrix confervicola.
    2.   ,,   Mucor.
    3.   ,,   luteola.
    4.   ,,   scopulorum.
    5.   ,,   fasciculata.
    6.   ,,   pannosa.
    7.   ,,   hydnoides,
    8.   ,,   cæspitula.

CIII. LYNGBYA.

Filaments elongate, decumbent, flaccid.

    1. Lyngbya majuscula.
    2.   ,,   ferruginea.
    3.   ,,   Carmichaelii.
    4.   ,,   speciosa.
    5.   ,,   flacca.

CIV. Microcoleus.

Filaments minute, rigid, straight, several enclosed together in a membranous sheath.

1. Microcoleus anguiformis.

CV. Oscillatoria.

Filaments needle-shaped, straight or slightly curved, short. Named from the curious motion observed in the filaments, which resembles the oscillations of a pendulum.

1. Oscillatoria littoralis.
2.   ,,   subsalsa.
3.   ,,   spiralis.
4.   ,,   nigro-viridis.
5.   ,,   subuliformis.
6.   ,,   insignis.

CVI. Spirulina.

Filaments spirally twisted, lying in a mucous stratum, vividly oscillating.

1. Spirulina tenuissima.
2.   ,,   Hutchinsiæ.

Order XVIII. NOSTOCHACEÆ.

Green, fresh-water, or rarely marine Algæ, composed of moniliform filaments, lying in a gelatinous matrix.

CVII. MONORMIA.

A single filament enclosed in a convoluted, gelatinous, and branching frond.

1. Monormia intricata.

CVIII. SPHÆROZYGA.

Filaments free, separate, naked.

1. Sphærozyga Carmichaelii.
2.   „   Thwaitesii.
3.   „   Broomei.
4.   „   Berkeleyana.
5.   „   Ralfsii.

CIX. SPERMOSEIRA.

Filaments free, separate, each enclosed in a very delicate membranous tube.

1. Spermoseira littorea.
2.   ,,   Harveyana.

Order XIX. PALMELLACEÆ.

Cells contained in confervoid, simple or branched, tubular filaments.

CX. HORMOSPORA.

Filaments gelatinous, confervoid, each enclosing a linear series of oval or spherical cells.

1. Hormospora ramosa.

ALPHABETICAL INDEX

OF THE

BRITISH MARINE ALGÆ:

LOCALITIES WHERE FOUND,

AND

TIME OF APPEARANCE.

ALARIA esculenta. Winter and spring. Falmouth. Shores of Scotland. North and West of Ireland.

Arthrocladia villosa. Summer and autumn. Mount Edgecumbe; Plymouth. Frith of Forth. Wicklow.

Asperococcus compressus. Summer. Mill Bay; Plymouth; Torquay.

Asperococcus echinatus. Summer and autumn. Redding Point; Plymouth; Torquay.

Asperococcus Turneri. Summer. Whitsand Bay; Devon; coast of Sussex. Bantry Bay.

BANGIA carnea. Spring. Glamorganshire.

Bangia ceramicola. Spring. Appin, Scotland.

Bangia ciliaris. Spring. Appin, Scotland. On the old leaves of Zostera marina.

Bangia elegans. Spring. Strangford Lough; Porta-
ferry. Parasitical on the smaller Algæ.

Bangia fusco-purpurea. Spring. Plymouth; Torquay.
West of Ireland.

Bonnemaisonia asparagoides. Summer. Mount Edgecumbe; Torpoint; Plymouth; Torquay. Salcoats;
Bantry Bay.

Bostrychia scorpioides. Summer and autumn. St.
Germain's River; Carbeal Mill; Plymouth; mouth
of the Dart, &c., &c.

Bryopsis hypnoides. Summer and autumn. Redding
Point; Plymouth; Torquay. Portrush, Ireland.
Scotland.

Bryopsis plumosa. Summer and autumn. Beggar's
Island; Torpoint; Plymouth; Torquay. Appin,
&c., &c.

Calithamnion affine. Summer. Shores of Bute, on
Fuci.

Calithamnion arbuscula. Autumn. Mewstone; Firestone Bay; Plymouth. Western shores of Scotland.

Calithamnion barbatum. Autumn. Ilfracombe; quay
at Penzance; Weymouth.

Calithamnion Borreri. Spring and autumn. Beggar's
Island; Torpoint; Plymouth; Brighton; Torquay;
Yarmouth.

Calithamnion brachiatum. Autumn. Bovisand; Plymouth. On Fuci.

Calithamnion Brodiæi. Summer. Torquay; coast of Northumberland; Cornwall; Miltown Malbay.

Calithamnion byssoideum. Spring. Mount Edgecumbe; St. Germain's River; Plymouth; Torquay.

Calithamnion corymbosum. Spring. Mount Edgecumbe; Firestone Bay; Plymouth; Torquay.

Calithamnion cruciatum. Spring. Firestone Bay; Torpoint; Plymouth; Torquay; Miltown Malbay.

Calithamnion Daviesii. Summer. Plymouth; Torquay.

Calithamnion fasciculatum. Summer. Yarmouth.

Calithamnion floccosum. Spring. Orkney Islands; Aberdeen.

Calithamnion floridulum. March and April. Land's End. Antrim. Orkney Islands.

Calithamnion gracillimum. Spring. Mount Edgecumbe; Beggar's Island; Plymouth; Torquay; Falmouth.

Calithamnion Hookeri. Spring and summer. Mount Edgecumbe; under the Hoe; Plymouth; Orkney. Ireland.

Calithamnion mesocarpum. Summer. Appin.

Calithamnion pedicellatum. Summer. Firestone Bay; Plymouth; Torpoint; Brighton; Torquay; Falmouth; &c.

Calithamnion pluma. Summer. Bantry Bay. Malbay. Appin.

Calithamnion plumula. Spring and summer. Firestone Bay; Mount Edgecumbe; Plymouth; Torquay; Falmouth.

Calithamnion polyspermum. Spring. Growing on Fuci. Mount Edgecumbe; Plymouth; Torquay; Penzance.

Calithamnion purpurascens. Spring. Brighton.

Calithamnion roseum. Spring and summer. Beggar's Island; Mount Edgecumbe; Plymouth; Torquay. Bantry Bay.

Calithamnion Rothii. Autumn. Bovisand; Plymouth; Mount Edgecumbe; Torquay.

Calithamnion sparsum. Autumn. On Laminaria saccharina. Appin.

Calithamnion spongiosum. Spring and summer. Firestone Bay; under the Hoe; Plymouth; Torquay; Salcombe.

Calithamnion tetragonum. Summer. On Fuci. Bovisand; Plymouth; Whitsand Bay; Torquay.

Calithamnion tetricum. Summer and autumn. Bovisand; Plymouth; Whitsand Bay; Torquay; Salcombe.

Calithamnion thuyoideum. Spring. Mount Edgecumbe; Torquay; Yarmouth. Wicklow.

Calithamnion tripinnatum. May. Roundstone Bay.

Calithamnion Turneri. Summer. Parasitical on other
Algæ. Bovisand; Whitsand Bay; Torquay.

Calithamnion virgatulum. Summer. Torquay, growing on Ceramium rubrum.

Calothrix cæspitula. Summer. On rocks near high-
water mark. Miltown Malbay.

Calothrix confervicola. Autumn. Mount Edgecumbe;
Torquay. Growing on decayed Ceramium.

Calothrix fasciculata. Autumn. On rocks below high-
water mark. Miltown Malbay.

Calothrix hydnoides. Summer. Appin.

Calothrix luteola. Summer. On marine filiform Algæ.
Appin.

Calothrix Mucor. Summer. On marine Algæ. Brighton.

Calothrix pannosa. Summer. On rocks near high-
water mark. Roundstone Bay.

Calothrix scopulorum. Autumn. On rocks near high-
water mark. Roundstone Bay.

Calothrix semiplena. Summer. On Corallina officinalis. Kilkee. Sidmouth.

Carpomitra Cabreræ. Winter. Mount Edgecumbe;
Torpoint; dockyard, Plymouth.

Catenella Opuntia. Summer. Mount Edgecumbe;
Bovisand. On rocks near high-water mark.

Ceramium acanthonotum. Spring and summer. Firestone Bay; under the Hoe; Plymouth; Torquay.

Ceramium botryocarpum. Summer. Mount Edgecumbe ; Torquay. Ardrossan.

Ceramium ciliatum. Summer. Firestone Bay; Bovisand ; Torquay ; Sidmouth.

Ceramium decurrens. Summer. Mount Edgecumbe ;
Bovisand ; Torquay.

Ceramium Deslongchampii. Summer. Mount Edgecumbe ; Torpoint ; Torquay ; Ilfracombe.

Ceramium diaphanum. Summer and autumn. Bovisand ; Whitsand Bay ; Torquay ; Sidmouth.

Ceramium echionotum. Summer and autumn. Bovisand ; Redding Point ; Plymouth ; Torquay.

Ceramium fastigiatum. Summer. Mount Edgecumbe ; Torpoint ; Torquay. Frith of Forth.

Ceramium flabelligerum. Summer and autumn.
Mount Edgecumbe ; Bovisand ; Whitsand Bay ;
Torbay ; Jersey.

Ceramium gracillimum. Spring and summer.
Mount Edgecumbe ; Beggar's Island ; Plymouth ;
Torquay.

Ceramium nodosum. Summer. Mount Edgecumbe ;
Bovisand ; Whitsand Bay.

Ceramium rubrum. Throughout the year. Very common everywhere, growing on rocks and Fuci.

Ceramium strictum. Summer. Under the Hoe ; Bovisand ; Whitsand Bay ; Torquay. Roundstone,

Chondrus crispus. Summer and autumn. Common in England, Scotland, and Ireland.

Chondrus Norvegicus. Autumn. Redding Point; Plymouth; Whitsand Bay; Bovisand.

Chorda Filum. Summer. Common everywhere.

Chorda lomentaria. Summer. On the mooring-buoys, Plymouth Harbour; Torquay.

Chordaria divaricata. Autumn. Carrickfergus.

Chordaria flagelliformis. Autumn. Bovisand; Cawsand Bay; Torquay.

Chrysymenia clavellosa. Spring and summer. Mount Edgecumbe; Firestone Bay; Plymouth; Torquay.

Chrysymenia rosea. Summer. Skaill, Orkney.

Chylocladia articulata. Spring and summer. Redding Point; Plymouth; Bovisand; Torquay.

Chylocladia kaliformis. Spring, summer, and autumn. Mount Edgecumbe; Bovisand; Whitsand Bay; Torquay.

Chylocladia ovalis. Spring and summer. Firestone Bay; Bovisand; Redding Point; Torquay; &c.

Chylocladia parvula. Summer and autumn. Firestone Bay; Cawsand Bay; Bovisand; Torquay.

Chylocladia reflexa. Autumn. Under the Hoe; Plymouth.

Cladophora albida. Autumn. On the larger Algæ. Bovisand; Cawsand Bay; Torquay.

Cladophora arcta.   Summer.   Firestone Bay;   Plymouth; Whitsand Bay.

Cladophora Balliana.   Summer.   Clontarf.

Cladophora Brownii.   Summer.   Land's End, Cornwall; Torquay.   Wicklow; Dunree.

Cladophora diffusa.   Summer.   Mount Edgecumbe; Torquay.   Portrush, Ireland.   Sidmouth.

Cladophora falcata.   Summer.   Dingle Harbour, Kerry. Jersey.

Cladophora flavescens.   Summer.   In ditches or pools of brackish water.   Common.

Cladophora flexuosa.   Summer.   Yarmouth; Torquay. Ballycastle.

Cladophora fracta.   Summer.   In ditches and pools of brackish water.   Common.

Cladophora Gattyæ.   Summer.   On rocks near low-water mark.

Cladophora glaucescens.   Summer.   Torquay; Falmouth Bay; Mount's Bay.   Portmarnock.

Cladophora gracilis.   Summer.   Torquay; Firestone Bay.   Youghal; Belfast Bay.

Cladophora Hutchinsiæ.   Summer.   Tor Abbey; Saltcoats.   Bantry Bay; Belfast Bay.

Cladophora lætevirens.   Summer.   Torpoint; Whitsand Bay; Torquay; Falmouth.

Cladophora lanosa.   Summer.   Torquay; Bovisand. On the larger Fuci.

Cladophora Macallana. Summer. Dredged in Roundstone Bay.

Cladophora Magdalenæ. Summer. Jersey.

Cladophora nuda.. Summer. On basalt-rocks, Port Stewart, Co. Antrim.

Cladophora pellucida. Spring and summer. On rocks. Firestone Bay; Mount Edgecumbe; Yarmouth. Belfast Lough.

Cladophora rectangularis. Summer. Torquay; Whitsand Bay. Galway; Roundstone Bay.

Cladophora refracta. Summer. Whitsand Bay; Cawsand Bay; Torquay; Ilfracombe. Kilkee.

Cladophora repens. Summer. Jersey.

Cladophora Rudolphiana. Summer. Parasitical on Zostera. Roundstone Bay. Falmouth.

Cladophora rupestris. Summer and autumn. Bovisand; Mount Batten; Whitsand Bay; Torquay. Connemara.

Cladophora uncialis. Summer. Whitsand Bay; Bovisand; Torquay. Rathlin; Antrim.

Cladostephus spongiosus. Summer. Firestone Bay; Bovisand; Torquay; Falmouth.

Cladostephus verticillatus. Summer. Whitsand Bay; Bovisand; Torquay; Sidmouth.

Codium adhærens. Summer. Torquay; Falmouth Harbour; Land's End. Rathlin; Antrim.

H

Codium amphibium. Summer. Roundstone Bay; coast of Galway.

Codium Bursa. Summer. Near Torquay; coast of Sussex; Cornwall. Near Belfast.

Codium tomentosum. Spring and summer. Firestone Bay; Bovisand; Torquay. Growing on Fuci.

Conferva ærea. Summer. Whitsand Bay; Torquay.

Conferva arenicola. Summer. Salt-marshes within reach of the tide.

Conferva arenosa. Summer. On the sandy shore, at half-tide level. Appin; Bantry Bay.

Conferva Bangioides. Summer. Breakwater, Plymouth; Torquay. Ballycotton.

Conferva clandestina. Summer. Under side of stones impregnated with putrefying marine substances. Weymouth.

Conferva collabens. Summer. Yarmouth.

Conferva implexa. Summer. Torquay. Bantry Bay; Malbay. Frith of Forth.

Conferva Linum. Summer. In the salt-water ditches near Dublin.

Conferva littorea. Summer. In the salt-water ditches near the muddy sea-shore.

Conferva Melagonium. Autumn. Whitsand Bay. Orkney. Cornwall.

Conferva sutoria. Autumn. In salt-water ditches and pools.

Conferva tortuosa. Summer. Torquay. Bantry. Berwick. Frith of Forth.

Conferva Youngana. Summer. Yarmouth. Dunraven Castle ; Dingle Bay ; Kerry.

Corallina elongata. Spring and summer. Coast of Cornwall ; Jersey.

Corallina officinalis, Throughout the year. Very common everywhere on rocky shores.

Corallina squamata. Summer. Whitsand Bay. South and West of Ireland. Jersey.

Crouania attenuata. Autumn and winter. Firestone Bay ; Plymouth ; Salcombe ; Land's End.

Cruoria pellita. Spring. Common on the rocky shores of Britain.

Cutleria multifida. Summer. Mount Edgecumbe ; Beggar's Island ; Plymouth ; Torquay.

Cystoseira barbata. Summer. Devonshire coast (doubtful).

Cystoseira ericoides. Summer and autumn. Redding Point ; Bovisand ; Whitsand Bay ; Torquay. Ireland.

Cystoseira fibrosa. Summer and autumn. Bovisand ; Whitsand Bay ; Torquay. Ireland.

Cystoseira fœniculacea. Summer. Bovisand ; Sidmouth and Torquay ; Weymouth ; Isle of Wight

Cystoseira granulata. Summer. Aberfraw. Torquay ; Bovisand ; Jersey.

DASYA arbuscula. Summer and autumn. Firestone Bay; Mewstone. Scotland. Bantry Bay.

Dasya coccinea. Summer and autumn. Mount Edgecumbe; Redding Point; Torquay; Sidmouth.

Dasya ocellata. Summer. Torpoint; Exmouth; Torquay. Wicklow; Balriggan.

Dasya venusta. Summer. Jersey.

Delesseria alata. Summer and autumn. Bovisand; Redding Point; Whitsand Bay; Torquay.

Delesseria angustissima. Winter and spring. North of Scotland. Yarmouth.

Delesseria Hypoglossum. Spring and summer. Mount Edgecumbe; Mount Batten; Plymouth; Torquay; Sidmouth.

Delesseria ruscifolia. Spring and summer. Mount Batten; Mount Edgecumbe; Plymouth; Torquay; Sidmouth.

Delesseria sanguinea. Spring and summer. Mount Batten; Mount Edgecumbe; Torquay; Exmouth. Saltcoats.

Delesseria sinuosa. Autumn. Bovisand; Redding Point; Plymouth; Torquay; Exmouth.

Desmarestia aculeata. Spring and summer. Very common on the shores of England, Scotland, and Ireland.

Desmarestia ligulata. Spring and summer. Mount Edgecumbe; Whitsand Bay; Torquay; Sidmouth.

Desmarestia viridis. Spring and summer. Mount Edgecumbe; Firestone Bay; Torquay; Exmouth.

Dictyosiphon fœniculaceus. Summer. Torpoint; Torquay; Exmouth; Sidmouth.

Dictyota dichotoma. Summer and autumn. Common everywhere.

Dudresnia coccinea. Summer. Mount Edgecumbe; Torpoint; Torquay; Exmouth.

Dudresnia divaricata. Summer. Bovisand; Whitsand Bay; Torquay.

Dumontia filiformis. Spring and summer. Firestone Bay; Torpoint; Torquay.

Ectocarpus amphibius. Summer. In muddy ditches of brackish water. Bristol.

Ectocarpus brachiatus. Summer. Mount Edgecumbe; Whitsand Bay; Torquay. Youghal.

Ectocarpus crinitus. Summer. Appin. Watermouth; Devon.

Ectocarpus distortus. Summer. Parasitical on Zostera marina. Appin. Saltcoats.

Ectocarpus fasciculatus. Summer. Whitsand Bay; Torquay.

Ectocarpus fenestratus. Summer. Salcombe.

Ectocarpus granulosus. Spring and summer. Firestone Bay; Whitsand Bay; Torquay.

Ectocarpus Hincksii. Summer. Plymouth; Torbay; Torquay. Aberdeen. Mount's Bay.

Ectocarpus Landsburgii. Summer. Dredged in deep water. Isle of Arran. Roundstone Bay.

Ectocarpus littoralis. Spring and summer. Abundant everywhere.

Ectocarpus longifructus. Summer. Parasitical on Algæ. Skaill, Orkney.

Ectocarpus Mertensii. Spring. Firestone Bay; Whitsand Bay; Salcombe; Torbay; Ilfracombe.

Ectocarpus pusillus. Summer and autumn. Parasitical on the smaller Algæ. Plymouth; Land's End; Torquay.

Ectocarpus siliculosus. Spring and autumn. Parasitical on other Algæ. Plymouth; Torquay; Jersey.

Ectocarpus sphærophorus. Spring. Parasitical on other Algæ. Plymouth; Torquay; &c.

Ectocarpus tomentosus. Spring. Parasitical on Fuci. Firestone Bay; Mount Edgecumbe; Torquay; &c.

Elachistea curta. Summer. On Fuci. Swansea.

Elachistea flaccida. Summer and autumn. Parasitical on Cystoseira fibrosa. Plymouth; Torquay.

Elachistea fucicola. Summer and autumn. Parasitical on Fuci.

Elachistea pulvinata. Summer and autumn. Parasitical on Cystoseira ericoides.

Elachistea scutulata. Summer. Parasitical on Himanthalia Lorea.

Elachistea stellulata. Summer. Parasitical on Dictyota dichotoma. Torquay.

Elachistea velutina. Summer. Parasitical on Himanthalia Lorea.

Enteromorpha clathrata. Summer. Whitsand Bay; Torquay; Exmouth.

Enteromorpha compressa. Spring and summer. Abundant everywhere.

Enteromorpha Cornucopiæ. Summer. On corallines, &c., in rocky pools left by the tide. Orkney.

Enteromorpha erecta. Summer. On rocks in the sea. Not uncommon.

Enteromorpha intestinalis. Summer. Very common.

Enteromorpha Hopkirkii. Summer and autumn. Torbay. Carrickfergus.

Enteromorpha Linkiana. Summer. Appin.

Enteromorpha percursa. Spring and summer. Tor Abbey. Appin. Clontarf. And on muddy seashores.

Enteromorpha Ralfsii. Summer. Bangor, North Wales.

Enteromorpha ramulosa. Spring. Rocks and stones between tide-marks.

Fucus canaliculatus. Summer and autumn. On rocky sea-shores. Abundant everywhere.

Fucus ceranoides. Spring and summer. Torquay; Exmouth; Torbay; Orkney; Falmouth.

Fucus Mackaii. April and May. Connemara; Loch Coul. East coast of Skye, &c.

Fucus nodosus. Winter and spring. Growing on sub-marine rocks. Abundant everywhere.

Fucus serratus. Winter and spring. Very common.

Fucus vesiculosus. Winter and spring. Abundant everywhere on rocks and stones at low water.

Furcellaria fastigiata. Summer. Redding Point; Firestone Bay; Torquay.

GELIDIUM cartilagineum. Summer. Freshwater Bay, Isle of Wight.

Gelidium corneum. Summer and autumn. On rocky shores. Very common.

Gigartina acicularis. Autumn and winter. Bovisand; Redding Point; Torquay. Belfast Bay.

Gigartina mamillosa. Autumn and winter. On rocks and stones near low-water mark. Common.

Gigartina pistillata. Spring and autumn. On rocks at extreme low-water mark. Whitsand Bay.

Gigartina Tedii. Autumn. Tor-Abbey rocks; Elberry Cove.

Ginnania furcellata. . Summer. Bovisand; Whitsand Bay; Torquay; Exmouth.

Gloiosiphonia capillaris. Summer. Redding Point; Plymouth; Torquay.

Gracilaria compressa. August. Sidmouth; [Exmouth. Thrown up from deep water.

Gracilaria confervoides. Summer and autumn. Torpoint; Firestone Bay; Torquay; Exmouth.

Gracilaria erecta.. Autumn and winter. Torquay; Sidmouth; Torpoint. Belfast Bay; Roundstone.

Gracilaria multipartita. Autumn and winter. Mount Edgecumbe; Torpoint; Whitsand Bay; Salcombe.

Grateloupia filicina. Autumn and winter. Whitsand Bay; Sidmouth; Ilfracombe; Torbay; Land's End.

Griffithsia barbata.. Summer. Brighton; Jersey.

Griffithsia corallina. Summer. Firestone Bay; Whitsand Bay; Torquay; Exmouth.

Griffithsia Devoniensis. Summer. Torpoint; Beggar's Island; Mount Edgecumbe (Plymouth).

Griffithsia equisetifolia. Summer. Mewstone, abundant; Whitsand Bay; Bovisand.

Griffithsia secundiflora. Spring and autumn. Bovisand; near Plymouth.

Griffithsia setacea. Spring and summer. Mount Edgecumbe; under the Hoe; Plymouth; Torquay; Exmouth.

Griffithsia simplicifolia. Autumn. Yarmouth. Black Castle, county of Wicklow.

Gymnogongrus Griffithsiæ. Autumn and winter. Cawsand Bay. Bantry Bay; Balriggan.

Gymnogongrus plicatus. Summer and autumn. Redding Point; Bovisand; Torquay; Sidmouth; Falmouth.

Halidrys siliquosa. Autumn and winter. Common. Found amongst rejectamenta.

Haliseris polypodioides. Spring and summer. Whitsand Bay; Mount Edgecumbe; Exmouth; Torquay.

Halymenia ligulata. Summer. Mount Edgecumbe; Bovisand; Torquay; Sidmouth.

Hapaladium Phyllactidium.

Hildebrandtia rubra. Summer. Torquay; Plymouth; Sidmouth.

Himanthalia Lorea. Spring and summer. Abundant everywhere.

Hormospora ramosa. Summer. Growing in a saltwater lake near Wareham, Dorset.

Hypnæa purpurascens. Summer and autumn. Redding Point; Bovisand; Torquay; Sidmouth.

Iridæa edulis. Summer and autumn. Bovisand; Firestone Bay; Torquay; Sidmouth; Falmouth.

JANIA corniculata. Summer. On smaller Algæ, between tide-marks. Plymouth; Jersey.

Jania rubens. Summer. On smaller Algæ, between tide-marks, on the south coast.

KALYMENIA Dubyi. Spring and summer. Firestone Bay; Mount Batten; Plymouth.

Kalymenia reniformis. Summer. Firestone Bay; Mount Batten; Sidmouth; Torquay.

LAMINARIA bulbosa. Autumn. On rocks, at low-water mark. Abundant on the British shores.

Laminaria digitata. Summer and autumn. Abundant everywhere.

Laminaria Fascia. Spring and summer. Mount Edgecumbe; Whitsand Bay; Torquay; Salcombe; Mount's Bay.

Laminaria longicruris. Summer. Ayrshire coast; coast of Banffshire; Dunluce Castle. Antrim.

Laminaria Phyllitis. Summer. Sidmouth; Torquay. Androssan; Bantry Bay.

Laminaria saccharina. Summer. Very common all round the coast.

Laurencia cæspitosa. Summer. Common on the shores of the British Islands.

Laurencia dasyphylla. Summer. Bovisand; Mount Edgecumbe; Torquay; Sidmouth; Exmouth.

Laurencia obtusa. Summer. Whitsand Bay; Redding Point. Androssan; Arran.

Laurencia pinnatifida. Spring and summer. Mount Edgecumbe; Bovisand; Torquay; Sidmouth; Exmouth.

Laurencia tenuissima. Summer. Bovisand; Torquay; Weymouth; Isle of Wight. Ballycotton.

Leathesia Berkeleyi. Summer. On submarine rocks between tide-marks.

Leathesia tuberiformis. Summer and autumn. Bovisand; Mount Batten; Torquay; Exmouth.

Lithocystis Allmanni. Summer. Parasitical on Chrysimenia clavellosa. Malahide.

Litosiphon Laminariæ. Summer. Parasitical on Alaria esculenta.

Litosiphon pusillus. Summer. Parasitical on Chorda Filum. Very abundant.

Lyngbya Carmichaelii. Summer. On marine Fuci. Plymouth; Torbay; Cornwall. Appin.

Lyngbya Cutleriæ. Spring and summer. Near the mouth of the Otter; Budleigh Salterton.

Lyngbya ferruginea. Summer. In mud-bottomed pools of brackish water. Appin.

Lyngbya flacca. Summer. Parasitical on various Algæ; on the Fuci.

Lyngbya majuscula. Summer and autumn. Mount

Edgecumbe; Torbay; Ilfracombe; Mount's Bay.
Belfast Bay.

Lyngbya speciosa. Summer. On marine rocks between
tide-marks. Torbay ; St. Michael's Mount.

MELOBESIA agariciformis. Summer. Roundstone Bay;
Connemara.

Melobesia calcarea. Summer. On the south coast of
England, and west of Scotland and Ireland.

Melobesia farinosa. Summer. On various Algæ.

Melobesia fasciculata. Summer. Lying on the sandy
bottom of the sea. Roundstone Bay.

Melobesia lichenoides. Summer. Coast of Cornwall.
Galway and Clare ; coast of Cork.

Melobesia membranacea. Summer. Common on the
leaves of Zostera.

Melobesia polymorpha. Summer. On rocks all round
the coast.

Melobesia pustulata. Summer. On Phyllophora rubens, and other Algæ.

Melobesia verrucata. Summer. On rocks round the
coast.

Mesogloia Griffithsiana. Summer. Between tide-
marks. South of England. West of Ireland.

Mesogloia vermicularis. Summer. Mewstone. Bovisand ; Torquay; Exmouth.

Mesogloia virescens. Summer. Whitsand Bay ; Redding Point ; Torquay ; Exmouth.

Microcladia glandulosa. Spring and summer. Under the Hoe ; Cawsand ; Plymouth ; Torquay.

Microcoleus anguiformis. Summer. Pools of brackish water. Dolgelly.

Monormia intricata. Summer. In ditches near Gravesend ; Shirehampton, near Bristol.

Myrionema clavatum. Summer. On Hildebrandtia rubrum.

Myrionema Lechlancherii. Summer and autumn. On decaying fronds of Rhodymenia palmata.

Myrionema punctiforme. Summer. Parasitical on Ceramium rubrum.

Myrionema strangulans. Summer. Parasitical on Ulvæ and Enteromorphæ.

Myriotrichia clavæformis. Summer. Parasitical on Chorda Lomentaria. Very common.

Myriotrichia filiformis. Summer. Parasitical on Chorda Lomentaria, and Asperococcus.

Naccaria Wigghii. Summer. Mount Edgecumbe ; Bovisand ; Exmouth ; Torquay.

Nemaleon multifidum. Summer. Bovisand ; Whitsand Bay ; Torquay.

Nemaleon purpureum. Autumn. Whitsand Bay ; Exmouth (very fine).

Nitophyllum Bonnemaisoni. Summer. On the stems of Laminaria digitata. Torquay and Ilfracombe. Youghal. Bute.

Nitophyllum Gmelini. Summer. Under the Hoe; Beggar's Island; Mount Edgecumbe; Plymouth.

Nitophyllum Hilliæ. Summer. Under the Hoe; Mount Edgecumbe; Firestone Bay; Plymouth.

Nitophyllum laceratum. Summer and autumn. Bovisand; Mount Edgecumbe; Beggar's Island; Plymouth.

Nitophyllum punctatum. Summer. Mount Edgecumbe; Redding Point; Plymouth; Torquay and Sidmouth.

Nitophyllum versicolor. Summer. Ilfracombe; Minehead.

OCHLOCHÆTE Hystrix. Summer. On stems in a lake of brackish water, Wareham, Dorset.

Odonthalia dentata. Summer. On rocks in the sea. On the shores of the north of England, Scotland, and Ireland.

Oscillatoria insignis. Summer. In a brackish ditch, Shirehampton, Bristol.

Oscillatoria littoralis. Summer. With the preceding.

Oscillatoria nigro-viridis. Summer. With the preceding.

Oscillatoria spiralis. Summer. On rocks by the sea-side. Appin.

Oscillatoria subsalsa. Summer. Brighton.

Oscillatoria subuliformis. Summer. In a brackish ditch, Shirehampton, near Bristol.

Padina Pavonia. Summer. Exmouth; Sidmouth; Torquay.

Peyssonelia Dubyi. Summer. On old shells. On the shores of the British Islands.

Phyllophora Brodiæi. Summer and spring. Whitsand Bay; Bovisand. Scotland. Belfast Bay.

Phyllophora membranifolia. Winter and spring. On rocky shores between tide-marks. Frequent.

Phyllophora palmettoides. Winter and spring. Exmouth; coast of Cornwall.

Phyllophora rubens. Winter. Common on the rocky shores of England and Ireland.

Plocamium coccineum. Summer and autumn. Mount Edgecumbe; Bovisand; Whitsand Bay; Torquay; Exmouth.

Polyides rotundus. Autumn and winter. On stones and rocks in the sea. Very abundant.

Polysiphonia affinis. Summer. On rocks in the sea. Carnlough, near Glenarm.

Polysiphonia atro-rubescens. Summer and autumn. Redding Point; Bovisand; Whitsand Bay; Torquay.

Polysiphonia Brodiæi. Throughout the year. On rocks, and the larger Algæ, between tide-marks. Common.

Polysiphonia byssoides. Spring and summer. Mount Edgecumbe; Beggar's Island; Plymouth. Ayrshire. Bantry.

Polysiphonia Carmichaeliana. Spring. Parasitical on Desmarestia aculeata. Appin.

Polysiphonia elongata. Spring and summer. Redding Point; Bovisand; Torquay; Exmouth; &c. Common.

Polysiphonia elongella. Summer. Redding Point; Whitsand Bay; Torquay.

Polysiphonia fastigiata. Spring and summer. Growing on Fucus nodosus and others. Abundant.

Polysiphonia fibrata. Summer and autumn. Growing on Fuci. Whitsand Bay; Bovisand.

Polysiphonia fibrillosa. Summer. With the preceding.

Polysiphonia formosa. Autumn. Torpoint; Mount Edgecumbe; Torquay.

Polysiphonia furcellata. Summer. Sidmouth; dredged in Torbay. Carrickfergus.

Polysiphonia Grevillii. Summer. On the larger Algæ. Shores of Bute.

Polysiphonia Griffithsiana. Summer. Parasitical on Polyides rotundus. Torquay; Isle of Portland.

I

Polysiphonia nigrescens. Autumn and spring. Mount Edgecumbe ; Torpoint ; Sidmouth ; Exmouth ; Torquay.

Polysiphonia obscura. Summer. On marine rocks, at half-tide level. Sidmouth and Jersey.

Polysiphonia parasitica. Summer and autumn. Firestone Bay ; under the Hoe ; Plymouth ; Torquay. Arran.

Polysiphonia pulvinata. Summer. Ilfracombe ; Torbay ; Whitsand Bay ; Salcombe ; Saltcoats.

Polysiphonia Richardsoni. Summer. Dumfries.

Polysiphonia simulans. Summer. Whitsand Bay ; Bovisand ; Torquay. Orkney. Jersey.

Polysiphonia spinulosa. Summer. Whitsand Bay. Appin.

Polysiphonia stricta. Summer. In the sea, on sand-covered rocks.

Polysiphonia subulifera. Summer. Torquay ; Weymouth. Belfast Bay.

Polysiphonia urceolata. Spring and summer. Abundant.

Polysiphonia variegata. Summer and autumn. Plymouth Harbour ; Beggar's Island ; Mount Edgecumbe.

Polysiphonia violacea. Summer. Mount Edgecumbe ; Torquay.

Porphyra laciniata. Spring to autumn. Very common on rocks and stones.

Porphyra miniata. Summer. Appin.

Porphyra vulgaris. Spring to autumn. Very common on rocks and stones.

Ptilota plumosa. Summer and autumn. On the stems of Laminariæ. On the northern and western coasts of Great Britain and Ireland.

Ptilota sericea. Summer and autumn. On rocks. Bovisand; Mount Batten; Whitsand Bay; Torquay; Exmouth.

Punctaria latifolia. Spring. Firestone Bay; Sidmouth; Torquay. Near Belfast.

Punctaria plantaginea. Summer. Firestone Bay; Torquay; Exmouth; Ilfracombe; Falmouth.

Punctaria tenuissima. Summer. Firestone Bay, &c., growing on Zostera marina.

Pycnophycus tuberculatus. Summer and autumn. Devon and Cornwall. Common.

Ralfsia verrucosa. Winter. On rocks between tidemarks. Common.

Rhizoclonium Casparyi. Summer. On sand-covered rocks near high-water mark. Not uncommon.

Rhizoclonium riparium. Summer. With the preceding.

Rhodomela lycopodioides. Summer. On the stems of Laminaria digitata.

Rhodomela subfusca. Spring and summer. On stones in pools, at low-water mark. Abundant.

Rhodymenia bifida. Summer. Mount Edgecumbe; Whitsand Bay; Torquay; Exmouth.

Rhodymenia ciliata. Autumn and winter. Mount Edgecumbe; under the Hoe; Plymouth; Torquay; Exmouth.

Rhodymenia cristata. Summer. Parasitical on Laminaria digitata. Caithness.

Rhodymenia jubata. Summer. In rocky and gravelly pools. Very abundant.

Rhodymenia laciniata. Summer and autumn. On rocks and stones. Under the Hoe; Bovisand. Mewstone.

Rhodymenia palmata. Summer. On rocks, and on the stems of Laminaria digitata. Common.

Rhodymenia Palmetta. Summer and autumn. Bovisand; Whitsand Bay; Torquay. On the stems of Laminaria digitata.

Rivularia applanata. Summer. On rocks and stones between tide-marks.

Rivularia atra. Summer. On rocks and corallines.

Rivularia nitida. Summer and autumn. South of England and West of Ireland.

Rivularia plicata. Summer. On the rocky sea-shore, above high-water mark. Torquay. Ayremouth.

Rytiphlæa complanata. Summer. Under the Hoe; Plymouth. West of Ireland.

Rytiphlæa fruticulosa. Summer and autumn. Bovisand; Whitsand Bay; Torquay; Exmouth; &c.

Rytiphlæa pinastroides. Winter. On submarine rocks near low-water mark.

Rytiphlæa thuyoides. Autumn. Whitsand Bay; Bovisand; Torquay; Sidmouth.

SARGASSUM bacciferum. Occasionally cast on shore. Orkneys. (Gulf-weed).

Sargassum vulgare. With the preceding.

Schizosiphon Warreniæ. Summer. On rocks at high-water mark. Plymouth; Sidmouth; Falmouth.

Schizothrix Creswellii. Winter. On rocks near high-water mark. Sidmouth.

Seirospora Griffithsiana. Summer. Mount Edgecumbe; Firestone Bay; Plymouth; Torquay; Salcombe.

Spermoseira Harveyana. Summer. In muddy ditches. Shirehampton, near Bristol.

Spermoseira littorea. Summer. In muddy, brackish ditches. Barmouth; Dolgelly.

Sphacelaria cirrhosa. Autumn. On Desmarestia aculeata and other Algæ.

Sphacelaria filicina. Summer and autumn. Mewstone. Dredged near the breakwater, Plymouth; Torquay.

Sphacelaria fusca. Summer. Plymouth; Sidmouth; St. Michael's Mount.

Sphacelaria plumosa. Summer. Mount Batten; Torquay; Exmouth. Orkney.

Sphacelaria racemosa. Summer. Frith of Forth.

Sphacelaria radicans. Summer. Torquay. Bantry, &c.

Sphacelaria scoparia. Summer. Firestone Bay; Torquay; Bovisand. Frith of Forth.

Sphacelaria Sertularia. Summer. Parasitical on various Algæ. North and West of Ireland.

Sphærococcus coronopifolius. Summer and autumn. Mount Edgecumbe; Bovisand; Torquay. Belfast. Bute.

Sphærozyga Berkeleyana. Summer. Ditches near Shirehampton, Bristol.

Sphærozyga Broomei. Summer. With the preceding.

Sphærozyga Carmichaelii. Summer. With the preceding, and on decayed Algæ. Barmouth. Appin.

Sphærozyga Ralfsii. Summer. With the preceding.

Sphærozyga Thwaitesii. Summer. With the preceding. Dolgelly. Portbury, Somerset.

Spirulina Hutchinsiæ. Summer. With the preceding.

Spirulina tenuissima. Summer. With the preceding.

Sporochnus pedunculatus. Summer. Mount Edgecumbe; Torpoint; Torquay. Frith of Forth.

Spyridia filamentosa. Summer. Firestone Bay; Exmouth; Holy Head.

Striaria attenuata. Spring. Torpoint; Firestone Bay;
Torquay.

Stenogramme interrupta. Autumn and winter. Mount
Edgecumbe; Torpoint; Bovisand; Minehead.
Cork Harbour.

Stilophora Lyngbyæi. Summer. Mount Edgecumbe;
Torpoint. Scotland and Ireland.

Stilophora rhizodes. Summer. Mount Edgecumbe;
Beggar's Island; Torquay; Jersey.

Taonia atomaria. Summer. Mount Edgecumbe;
Whitsand Bay; Torquay. Ballycotton.

Ulva Lactuca. Autumn and summer. Whitsand Bay;
Bovisand.

Ulva latissima. From spring to winter. In the sea,
on rocks or stones. Very common.

Ulva Linza. Summer. With the preceding.

Vaucheria marina. Summer. Torbay; Salcombe.
Appin.

Vaucheria submarina. Summer. On the muddy sea-
shore, Weymouth.

Vaucheria velutina. Summer. On the muddy shore, Appin.

WRANGELIA multifida. Summer and autumn. Mount Edgecumbe; Redding Point; Mount Batten; Torquay. Belfast Bay.

ZONARIA collaris. Summer. Granville Bay; Jersey.
Zonaria parvula. Spring and summer. On rocks and corallines.

Printed by E. Newman, 9, Devonshire St., Bishopsgate, London.

Milton Keynes UK
Ingram Content Group UK Ltd.
UKHW011322120324
439381UK00009B/735